MISSIONS
TO
MARS

MISSIONS
TO
MARS

A NEW ERA OF ROVER AND SPACECRAFT DISCOVERY ON THE RED PLANET

Larry S. Crumpler

HARPER
DESIGN

An Imprint of HarperCollins Publishers

HarperCollins books may be purchased for educational, business, or sales promotional use. For information, please email the Special Markets Department at SPsales@harpercollins.com.

First published in 2021 by
Harper Design
An Imprint of HarperCollins*Publishers*
195 Broadway
New York, NY 10007
Tel: (212) 207-7000
Fax: (855) 746-6023
harperdesign@harpercollins.com
www.hc.com
Distributed throughout the world by
HarperCollins*Publishers*
195 Broadway
New York, NY 10007

ISBN 978-0-06-304736-5

Library of Congress Control Number: 2021025709

Printed in Malaysia

First Printing, 2021

DEDICATED TO JAYNE AUBELE,
my wife, friend, and fellow planetary geologist,
who was there through the whole journey.
As we have always said, together we have made
one smart person.

———————

"It is good to renew one's wonder," said the philosopher.
"Space travel has again made children of us all."
—RAY BRADBURY, *The Martian Chronicles*

CONTENTS

PART

1

KNOWING THE UNKNOWN

ONE

Full Circle

"The Kiowas reckoned their stature by
the distance they could see."

—N. SCOTT MOMADAY, *The Way to Rainy Mountain*

LOOKING AT EARTH FROM MARS

THE SCENE THAT LAY BEFORE US looked like the rubble from a crumbled, ancient city. We were standing out in the open on the surface in the frigid predawn glow, looking east in the midst of a plain littered with dark, dust-coated, angular rocks. We were surrounded by small craters, and the vista contained all the bleakness of an aftermath of some apocalyptic no-man's-land. Above the eastern horizon, pale in the predawn haze, was a small, blue-white star hanging silently in the sky. That small star was Earth, and we were seeing it from the surface of Mars.

With that glance we had come full circle. It was not a momentous occasion, judging from the casual interest we seemed to place on that observation, but there was an abiding sense that not only were we a long way from home but we were also a long way from our humble beginning in our quest to know the red planet. We were finally on the surface of the red "star" that humans had observed for such a long time from the yonder blue-white "star." As the day progressed, the Sun ascended from the alien blue of the strange dawn, and those of us looking at this scene virtually through the eyes of the Spirit rover continued our journey up the rocky slope, soldiering on to our next destination and another day's work on the red planet, Mars. The moment had passed, but it was the end of a beginning.

But the moment was actually epic in its meaning. For thousands of years, Mars had been a "red star" in the sky. Only recently in the chronicle of human exploration, in fact within just the last few decades, have we been in a position to do better in our quest for understanding Mars than the simple tracking of the motions of the "fiery star." Getting beyond this more primal association of the stars and mortal events, however, has been a difficult journey in the history of human interest in Mars. One of the more interesting ironies of the past decade of Mars exploration that illustrates how far we have come is that simple image taken early in the mission of Mars Exploration Rover Spirit. In the photo, we see the Martian predawn sky looking east, and in this sky is a tiny star, Earth. It is a subtle yet dramatic image, accompanied in the press release image by an arrow and the words "you are here." It shows our small place in the cosmic scheme, of course, but to me it is an image evoking how things have changed in our view of the planets, and Mars in particular. Here we were on Mars looking at a bright, bluish-white morning star that was another world, yet so far and

RIGHT: This was the view of Mars and the nearby Moon from Earth for centuries. This photograph of Mars in the evening sky was taken during the 2020 opposition. The view looks across the Rio Grande Valley from Rio Rancho toward the Albuquerque city lights.

BELOW: Sol (day) 63 image of the morning sky taken by Spirit rover one hour before sunrise. This is the first image of Earth taken by a spacecraft from the surface of Mars.

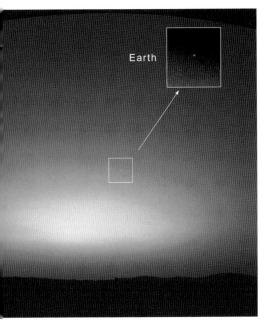

so small in the sky that, if you didn't already know all the teeming life and business of Earth, you could not tell much about the presence or absence of life on Earth from this tiny dot.

Across the distance that it takes light to travel in twenty minutes between Earth and Mars and forty-five years earlier I was one of those on that small, blue-white morning star viewing Mars from afar. It was during the twilight of telescopic Mars observation and the beginning of the space age that I began setting up my small telescope in the backyard. And whenever Mars was in the evening sky I attempted to look at it. I recall that on one particularly clear night I could see what I suspected was one of the polar caps, but not much else. All those dark markings and such were the province of observers with larger and professional telescopes. Mars was always a somewhat frustrating object in this respect. One was aware that there would probably be a lot of interesting things to see if you could just find a way to see them. Now, after all those years and all that wondering what it was like on that planet, I was one of those on the other end of that gulf between two planets, not only looking back at Earth, but looking at the surface of Mars up close and personal.

SEEING RED: MARS MANIA, THE EARLY YEARS

MAYBE THE INTEREST IN MARS IS simply a consequence of popular culture. Or who can say, maybe there is something else going on. There is the speculation among some people that life on Earth might easily have originated on Mars and was transferred here in the first few hundred million years of our planet's existence. That could explain the interest we humans have had in Mars; our interest has been the vague desire buried in our DNA to see the "home world" again. For now, that is only unsatisfying speculation and does nothing to solve the riddle of life's ultimate origin. Yet the search for the answer to that question—that is, whether there was ever life on Mars or whether it is present even today—is one that we have been pursuing for many decades. It is one of many stories that we will come to examine. But for now, there is no single explanation backed up by evidence or careful social analysis regarding why Mars has been such a focus of attention through the centuries.

Some of the attention on Mars from the beginning of recorded history undoubtedly reflected an interest traceable to the appeal that the color red has for most humans. Red sunsets, red deserts, red sports cars—you name it and if it is red it is probably visually more arresting for most people. Early cultures have had a fascination with Mars's perceived magical properties as a celestial object in its astrological role and in interactions with other celestial bodies, or its mythical origins and associations with war or other fiery events of mayhem that loom large in the human condition. And then there is a more modern obsession born of Mars's rich history in fictional and imaginative depictions as an abode of interesting events, creatures, or societies. And finally, the fascination that the lay public may have favors Mars as a new destination that is colorful, far away, and a sort of "new Southwest" with all

Some of the attention on Mars from the beginning of recorded history undoubtedly reflected an interest traceable to the appeal that the color red has for most humans.

manner of exotic and richly colored scenery to be explored and seen by someone bold enough to do so.

For a very long time, humanity has been in the position of looking at Mars from Earth as an oddly moving star, the "fire star," as the early Chinese observers called it. This was the initial way we looked at planets, not by standing on their surfaces or peering at them through telescopes, but by looking at them as tiny dots of light in the night sky that moved unlike the steady stars.

Probably in the time before recorded history, the red color of Mars was an obvious distinction because few objects in the night sky are so intensely ruddy to the unaided eye. The other thing that made Mars stand out to the eyes of ancient nighttime viewers was the fact that it was one of those special wandering stars too. This penchant Mars had in which it appeared to move backward from its normal motion for a period of time—referred to as retrograde motion—when combined with its unusual color helped to create its special place in astrological lore of the night sky, as recorded by ancient Egyptian astronomers more than four thousand years ago. The seemingly erratic changing of direction and the fiery appearance probably initiated thoughts of chaos, burning, and its destructive nature, inviting its association with the consequences of war.

Beyond these mythical associations, early observers were concerned about making sense of the timing of the return of objects and periods of an object's visibility rather than any interest in the objects themselves. The Chinese, for example, were interested in conjunctions of Mars—that is, the time when Mars goes behind the Sun—and kept a careful account of the motions of Mars even before the Zhou dynasty, more than three thousand years ago, if not before. And they had good reason to be concerned because accurate calendars, which were in part built around observations of the celestial motions, were important as part of the authority of the ruling dynasties. And another important reason for an interest in keeping a very accurate tracking of the planets was the fact that to them Mars was associated with all sorts of bad events and war. Keeping track of Mars was just good business.

The comings and goings of Mars in the night sky were important enough to early civilizations that the Mayans and Babylonians are known to have maintained a document of observations of Mars's movements. So while this interest can be traced back historically to the earliest known astronomical observations of the planet by the Chinese before 1000 BCE, and many civilizations since then, Mars was little more than a point of light to the human eye, and there was simply no way to know anything about it as an object. Mars was a mythical placeholder for deities in the sky and the clockwork machinations of the fates, but not much else.

THE TRICKSTER PLANET (COYOTE MARS)

IN MY QUEST TO UNDERSTAND HOW various groups had viewed Mars in the night sky over the years, I spoke with friend Joseph Aragon of the Acoma Pueblo in western New Mexico. I asked him what the Acomas called Mars. He told me that the traditional word for Mars meant simply "crazy star," perhaps a reflection of its back-and-forth motion in the sky, or the retrograde motion that the geometry between Earth's and Mars's orbits appears to impart on the progress of Mars across the sky.

Then I asked him about something else, a question that had emerged when I reflected on the twists and turns of the planet's exploration over all the years. As we will see, Mars has a habit of doing a kind of bait-and-switch on us as we learn new things. Just when we think we have gasped some new understanding, it often happens that the new knowledge is not the most important focus. Rather, there is something else that we are about to discover that is really the main point. When I mentioned this to Joe, I asked if Coyote, the eternal "trickster" so prominent in Native American legends, could in any way be similar to this penchant that Mars had for giving us little runarounds. He reflected for a minute and said that yes, it was probably entirely appropriate to think of Mars that way. After all, Coyote is also a wise figure but mischievous.

So, I began to think of Mars as "Coyote Mars" because there has always been this little bait-and-switch thing going on when we try to understand something. It is simply Coyote Mars doing what comes naturally. As Joe put it, Coyote in his wisdom was just "making us all dig a little deeper" to find the really meaningful thing behind something—in this case, the science of Mars.

TWO

Eyes on Mars

"It is possible to believe that all the past
is but the beginning of a beginning, and that all that is
and has been is but the twilight of the dawn."
—H. G. WELLS, *The Discovery of the Future*

A NEW WORLD VISIBLE
IN THE TELESCOPE

THE FOCUS ON WHAT MIGHT BE called the astrological angle changed during the Enlightenment, also known as the Age of Reason, a time of evolving ideas that developed in the mid-1600s through the end of the eighteenth century. This was a period when there was a movement to a new philosophy that advocated that observations and reasons could serve as the main methods of understanding the world rather than beliefs and portents. New and more powerful ways of observing were coming into existence. One of these was the telescope, and that naturally and very quickly was turned to the heavens. This was also a period when the Copernican revolution had taken hold and Earth was demoted from its centrality in the universe and became just one of many bodies circling the Sun. The idea that Earth was like the other planets and heavenly bodies no doubt inspired the foundation of ideas prevalent at the beginning of the Enlightenment, namely, that if Earth was a real world with inhabitants and environments, then maybe the other planets were too. All this was possibly aided by the fact that even here on Earth people were beginning to discover new worlds or continents. So why not new worlds in the heavens?

According to the then newly resurrected Aristarchian idea of a heliocentric solar system in the seventeenth century, Johannes Kepler and Nicolaus Copernicus had shown that if Mars were indeed orbiting the Sun beyond Earth's orbit, Mars should show gibbous-type phases like the Moon when it is between one-half and fully illuminated. Galileo was the first to turn a telescope on Mars in 1609, attempting to observe the phases of Mars, although his telescope was too crude to detect the phases. Nonetheless, in true scientific fashion, while he admitted that he could not say for sure whether the phases were present, he thought he could see that the illuminated disk was not entirely circular: an early example of the scientific method in which conclusions are born of observation, theory, more observation, and analysis. In other words, the observations could be explained as "consistent with" the shape of the disk if it were under gibbous-type illumination.

This was a time of rapid succession of many firsts in our understanding of Mars. Shortly after Galileo's first experimental observation, the first documentation of

Mars as a potential Earthlike body—perhaps even like Earth with seasons and land-masses, in other words a place with an actual geography—resulted from telescopic observations and drawings of Mars by Francesco Fontana of Naples, Italy, a lawyer and an astronomer, in 1636.

Unfortunately, the optics of his telescope were sketchy and the drawings he made are without detail regarding the planetary surface, but at least he was exploring the planet and recording what he thought he was seeing.

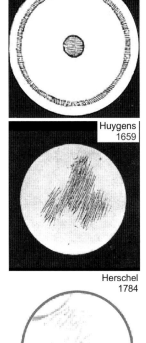

Fontana
1638

Huygens
1659

Herschel
1784

But the Dutch mathematician, astronomer, and telescope maker Christiaan Huygens added fuel to the fire of Mars interest with his observations of Mars through his 7-centimeter aperture 6.5-meter-long telescope.

Notes and sketches from his observations again adhere remarkably to the scientific method of the time, identifying surface markings on November 28, 1659, that were observed to move across the face of the Martian disk. This amazing observation led to the first estimates of the approximately twenty-four-hour rotation period of Mars. Even more remarkably, shortly afterward this estimate was refined to about twenty-four hours and forty minutes, very close to the now accepted length of the solar day on Mars (twenty-four hours thirty-nine minutes), a period of time that we in the era of spacecraft have come to know as a *sol* to distinguish it from an Earth (twenty-four-hour) day.

A long line of astronomers was making notes on their observations of Mars, including the Italian astronomer G. D. Cassini, who in 1666 noted the presence of white caps on the planet. Huygens further detected the white south polar cap in 1672, and with that the game was on to show ever more Earthlike characteristics. These were the luminaries of the sciences at the time, but there are many other examples of early observations of Mars by lesser-known telescopic observers that were remarkable for the level of detail and discipline of recording just the facts.

Beyond the philosophical explorations about the fact that planets were actual worlds in their own right, these first telescopic observations aroused interest among scientists in the exploration of planetary surfaces and, of course, the potential for life beyond Earth. This began a whole new philosophy, including the plurality-of-worlds line of thought that was made popular by Bernard le Bovier de Fontenelle in his 1686 fictional dialogues *Entretiens sur la*

ABOVE: Sketches from the earliest efforts to view Mars through telescopes.

pluralité des mondes. In his work, Fontenelle explores the possibility of life on other worlds from this new standpoint that considers that the planets, the Moon, and other objects in the night sky are objects like Earth circling the Sun according to the new Copernican view of things. The book is also famous for many other reasons. It was the first book about a scientific topic written in the popular language—in this case French, rather than Latin. Not only that, Fontenelle makes the fictional dialogue take place between a gentleman and a lady in a garden, thus inviting women for the first time to participate in the scientific discourses of the day, which up until then had been considered only an enterprise for men.

The British astronomer and famous telescope builder Sir William Herschel went on to make observations of Mars from 1777 to 1783.

While Herschel concentrated most of his activities on stellar observations, he took advantage of times when Mars was closest to Earth to identify several fundamental characteristics of Mars that could be determined with his telescope. He established a very accurate early estimate of the rotational period of Mars, determined that the atmosphere must be very thin based on the occultation of a star by Mars, and determined an axial inclination of Mars of around 24 degrees, a value similar but slightly greater than Earth's inclination. He also noted in his publications that Mars's polar caps varied seasonally. The seasonal nature of surface changes on Mars fired the imaginations of everyone because it invited the notion that Mars's seasons must be like those on Earth where there are winter snows and a spring thaw.

It is not surprising perhaps that Huygens was among the first scientists to entertain the notion that there was winter snow to form a polar cap and water following the seasonal thaws, and all this suggested an atmosphere. There was even the possibility of extraterrestrial life.

FIRST MAPS OF MARS

MARS CARTOGRAPHY STARTED WHEN JOHANN HEINRICH von Mädler and his student, a well-off banker, Wilhelm Beer, began to assemble a series of drawings of Mars into a global map in 1830. Thus was born the first true Mars map, the famed Beer-Mädler Mars map.

Because they had a map showing the fixed light and dark toned telescopic features of Mars, they were then in a position to have another go at estimating the

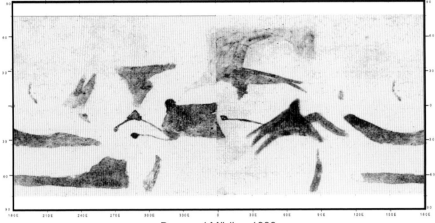

Beer and Mädler, 1830

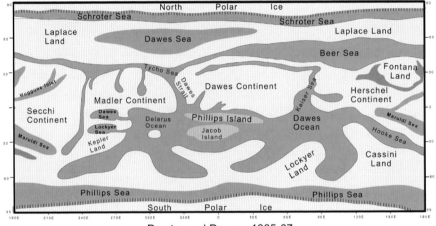

Proctor and Dawes, 1865-67

rotation period of Mars. After an early attempt with some inaccuracy, they arrived at a final estimate that was only about one second longer than the value we know today. In the process of this work, in 1832 they picked a spot as the prime meridian, later (1837–1841) named Sinus Meridiani by the French astronomer Camille Flammarion, the great dark area near the Martian equator, a sort of Greenwich of Mars, which is still used today with some precision cartographic tie points that were developed later. The exact location is now centered on a small crater, Airy-0.

Throughout the nineteenth century, things were ramping up in the early science of Mars cartography. In 1867 English astronomer Richard Proctor, taking advantage of several decades of advances in telescopes and methods, took a stab at Mars cartography, drafting a newer map of Martian features based on "a charming series of tracings supplied me by Mr. [William R.] Dawes."

Dawes was an accomplished amateur astronomer and had undertaken a series of observations and sketches of Mars from 1864 to 1865, which he had presented to the Royal Astronomical Society and had caught the eye of Proctor. No map is complete without names, so going further Proctor assigned names to Dawes's Martian map features. He used the names of prominent astronomers; it has been pointed out the names were somewhat skewed to British scientists, with Dawes's name appearing frequently. As Proctor related, "I have applied to the different features the names of those observers who have studied the physical peculiarities presented by Mars. Mr. Dawes' name naturally occurs more frequently than others. Indeed, if I had followed the rule of giving to each feature the name of its discoverer, Mr. Dawes's name would have occurred much more frequently than it actually does."

The naming scheme was interesting but was later overshadowed by a more "scientific" naming theme to be introduced by the famed Giovanni Schiaparelli and, further along, Eugène Antoniadi. But it was in these early maps that the dark areas became associated with seas and the lighter areas landmasses, a trend that followed through later maps even though it was to become less and less likely that there were actual seas. More to the point many of the dark "seas" were later shown to be vast uplands, and likewise many "continents" were low and bright with dust.

The presence of water was suspected very early though. The world of Mars mania experienced an explosive growth in 1871, when William Huggins used an optical device that splits light into a spectrum, a spectroscope that revealed the composition of the atmosphere of Mars for the first time. He announced the detection of water vapor. Richard Proctor noted that the presence of water vapor was certainly possible given the changing polar caps and possibility of melting snow and water ice that those changes suggested. He even went on to speculate about the life that the presence of water inevitably implies. He even expanded the speculation in an early bit of astrobiology hypothesis thinking to suggest that the lower gravity of Mars would support the development of Martians somewhat taller than humans and other benefits from the lower gravity. These ideas possibly planted seeds for the idea of strange alien biologies in general and strange Martians in particular. In a few years, however, as Proctor's theoretical review of Mars's conditions started to be more fully developed, he began moving toward a remarkably modern view of Mars, noting that given how cold Mars was, the fact that it was not entirely covered with white snow and ice implied that there was a lot less water vapor available and that the atmosphere was probably fairly thin. He concluded that the atmosphere was probably like being several kilometers up in Earth's atmosphere. And then to take a step even further forward into modern astrobiology, he noted that air that thin was probably unlikely to support life. Later still, he boldly predicted that Mars may have been like Earth in its early history, but it dried out and any life that may have existed died out. These thoughts, particularly the idea of an early wet Mars, are very close to those we believe today. In fact, it became a tenant of Mars speculation, including in science fiction, in years to follow: Mars was once rich with water, but as the climate dried out the Martians were left suffering, and maybe envious of Earth. The thread of fact and fiction is woven tightly in the history of thinking about Mars.

The comparison with Earth was carried forward in the interpretation of many of these early observations that the dark areas on Mars were likely seas and the orange patches were land. Enter actual scientific analysis of Mars at this time, again in the form of English astronomer Richard Proctor, who noted that given the relatively thin and probably colder air of Mars, seas were unrealistic and that any possibility of such an environment supporting life was sketchy at best. But the desire for a living Mars moved forward in the culture of the day, and in 1860 the French astronomer Emmanuel Liais proposed instead that the dark areas were old seabeds filled with vegetation. This notion persisted through the modern period of telescopic observation and in some respects into the early half of the twentieth century.

The year 1877 was a big year for Mars. It was the year of an "opposition," a point in Mars's orbit when it is closest to Earth. Many observers from many different nations were taking advantage of the opportunity for higher-resolution views of Mars through the best telescopes. Incidentally, it was during this opposition that Asaph Hall, an American astronomer, discovered and named the moons of Mars, Phobos and Deimos, the traditional companions of Mars as the god of war. It was during the same opposition of Mars in 1877 Schiaparelli, from Italy, kicked off the modern telescopic fascination with Mars by generating new charts and detailed drawings of the red planet. Not only did he make detailed drawings of the surface, he also suggested a new naming scheme for Mars's features, which departed from Proctor's style of using the names of individuals, primarily the names of astronomers who had, up to that time, contributed to observations of Mars. Instead he proposed that the regions of Mars could be named after places in Greek and Roman mythology, especially those related to Mars, and some biblical names. These names were later incorporated into more detailed Mars maps of the space age. Most of the features on his map were albedo features, bright and dark markings that would later be shown to have little connection with actual physical terrain features soon to be revealed with spacecraft. Mars was playing its tricks on us with what looked important, but in the end those features were marginalized and almost forgotten in a wave of the new technology's hand. However, as we will see later, some named features were resurrected and applied to Mars's physical terrain features.

More important, Schiaparelli's work inadvertently fostered the next big speculation about Mars that lasted for nearly the next one hundred years in one form or another. It was in these drawings he famously noted that the planet's deserts were crisscrossed with what he saw as many fine, often parallel lines, which he denoted, as channels, or *canali* in Italian. If you look at Schiaparelli's original map, the "lines" are not the ruler-straight lines of later maps, so he appears to have been careful about what he was drawing and not interpolating excessively. Either that or his sketching hand was a bit shaky. Schiaparelli was an observer and did not in any way speculate that the *canali* were artificial. Instead, the word *"canali"* was incorrectly translated into English as "canals." This was a classic case of what you did not say—and what people thought they heard you say—becoming more exciting than what you said. Hence was born the idea of Martian canals.

Into this fray then was cast the American astronomer Percival Lowell, who observed Mars through a 58-centimeter Alvan Clark telescope in his observatory atop a ridge in Flagstaff, Arizona, in the summer of 1894. It was at this time that the southern polar cap was diminishing and the apparent seasonal changes in color were manifesting. But his drawings and reports based on his observations generated considerable interest because they seemed to indicate, and were certainly interpreted as such by Lowell, that there was a complex network of what must be canals on Mars. It was not just that there were features that looked like canals, but their presence and the shifting patterns of light and dark across the Martian surface with the Martian seasons all seemed to fit a kind of pattern one might expect for vegetation being renewed by canal-channeled snowmelt over the far-flung surface of a parched planet.

There followed many attempts by other observers to see these lines on Mars, but no one else seemed able to see them. Eugène Antoniadi had become a noted Mars observer and was at first somewhat supportive of the canals but, following access to the large telescopes of Meudon Observatory in France during 1909, was unconvinced that they existed. His map of Mars published in 1929 in his book *La Planète Mars*,

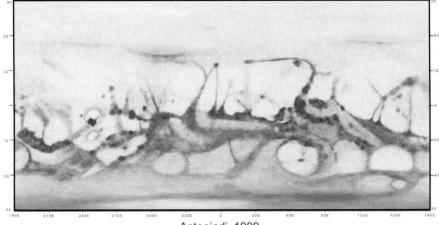

Antoniadi, 1909

the first of many books to bear that simple title, became the standard for Mars features right up to the very beginning of the space age.

Other observers who had spent considerable time before Lowell's reports and afterward viewing Mars offered a variety of explanations for the possibility that the lines were illusions of shapes and blotch boundaries and alignments of dark spots.

But many scientists simply began to question whether Lowell was just seeing illusions. One of his critics, Alfred Russel Wallace, was a scientist better known for other contributions in the natural sciences, such as the foundations of natural selection perhaps more widely championed by Charles Darwin. But he was also one of the first scientists to examine the potential for life elsewhere beyond Earth, or a subject that is known today as astrobiology. In his 1907 book *Is Mars Habitable?*, Wallace laid out the facts about Mars that he felt Lowell should have known better than anyone. The book reviewed the results of extensive research and collaboration with other experts about a variety of characteristics that one can quantitatively deduce about Mars that address the big questions about habitability. To summarize a very lengthy, and even today, interesting and instructive book, he basically stated that the atmosphere of Mars was known to be very thin; there was very little to no water on Mars, certainly no bodies of water from which to cycle the water into the atmosphere and onto the land and back to the atmosphere as is done on Earth; and the polar caps were likely some frozen heavier gas like carbon dioxide (CO_2) rather than water ice.

In any case, ever the gentleman, regarding Lowell's credibility as a scientist Wallace concluded, "I venture to think that his merit as one of our first astronomical observers will in no way be diminished by the rejection of his theory, and the substitution of one more in accordance with the actually observed facts."

And so there it was, the description of Mars a good sixty years before we determined that Mars was in reality a cold, barren desert. This was a good summary of what we confirmed as a profound observed fact after eventually going there. But the belief in a viable Mars kept the whole possibility of Martian life going, albeit lamely, until Wallace's assessment was shown to be right. Another lesson, and one that philosophers of science have noted from broad surveys to determine how widespread or generally valid the thought might be, is the fact that it is often the scientist from outside a specialty who has the clearest insight into some of a given specialty's more vexing problems.

Telescopic research on Mars continued through the early twentieth century and right up to the space age. By the early twentieth century a virtual roster of who's who in the planetary astronomy world had weighed in on observations of Mars. Few appeared to have any success seeing the strange lines on the surface of Mars that Schiaparelli, and, more vocally, Lowell, had allegedly documented, but interesting new results about the atmosphere and even the composition of the surface were being documented. Much of the work was beginning to zero in on the fact that Mars's atmosphere was thin, mostly carbon dioxide, and that there was very little water. In any case, the assessment that Mars was unlikely to fulfill any of the fantasies of life beyond Earth was insufficient to halt interest in understanding Mars as another planet. Throughout the early twentieth century serious progress was made with new and better telescopes as well as new instruments.

At this point in time it seems that a critical mass of scientific information combined with misunderstanding had accumulated such that imaginative storytelling was in a position to offer its take on the fascination of Mars. Even then the well-known H. G. Wells story *The War of the Worlds* about invading Martians was apparently prompted by a discussion he had with his brother Frank one day. They were discussing the disruption of the native populations by Europeans when his brother asked what would happen to our civilization should some "beings from another planet" come to Earth and begin "laying about them here!"

Obviously, Wells took that idea up as a story line, and from what better planet to have the beings originate than Mars and the very controversial proposed civilization of Mars proposed by Percival Lowell? A little current events story in the public media always goes a long way toward giving a fictional story gravitas. Of course, most science fiction fans have read at least some of Edgar Rice Burroughs's tales of John Carter of

In the case of Mars, despite the scientific conclusions that Mars was a cold, dry place, there had been a desire that it be like Earth.

Mars. While there was little in the Burroughs books about anything remotely related to the Mars known at the time, he did focus his stories on the fact that Mars was arid and that the civilization was coping with a barren environment. Beyond that, it was just a nice series of stories in which a mythical Mars was a setting. Nonetheless it kept the whole mystic of Mars in the public conscience.

One day when I actually sat down and did a binge reading of all the John Carter of Mars books, I came to realize that I had obliquely crossed paths with this series. I believe that it is the third book, *The Warlords of Mars*, that begins with John Carter at the Earth site from where he was able to go to Mars, which always took place as a kind of magical transference when he was in a dreamlike state. This transference to Mars happened in a cave that in this book was revealed as being in a volcanic landscape at the headwaters of the Little Colorado River in Arizona. Well, it so happens that during the late seventies through early eighties, I was actually engaged in geologic mapping of that very volcanic area. Although at the time I had not yet read the John Carter of Mars books, had I known about this magical cave I might have tried my hand at being transferred to Mars. In any event, the actual headwaters of the Little Colorado River are in the margins of the young pine-covered Springerville volcanic field near the little town of Greer. But I do not recall encountering any caves there.

The crossing of paths with some element of the Mars myth might appear unlikely, but, in reality, so much had been written in fiction that the chances of interacting with something that had been touched by one of those Mars stories is probably greater than one might otherwise assume. The Burroughs books were just one of the modern cultural elements of the broader fascination with Mars brought about no doubt by the speculations that there could be life on Mars and maybe even entire civilizations.

To summarize the history up to this point, the fascinating and instructive thing about this quest to determine whether Mars was Earthlike or not is the realization that in several instances the correct answer was deduced from the then available observations, namely that the conditions on Mars are those we know today: Mars is intensely cold and very dry, and has a very thin atmosphere of carbon dioxide. This was not the first time that a correct answer was proposed for some characteristic of the solar system, in this case for Mars, yet not adequately emphasized in science discussions that followed for many years. For example, Aristarchus had proposed the heliocentric model of the solar system as early as around 300 BCE, a model that proposed that the Sun was the center around which all the planets, including Earth, revolved. But this was an idea that was not accepted until nearly two thousand years later. Sometimes we have "the correct answer" but refuse to believe it, perhaps because the proposed answer is not what we have grown accustomed to thinking as the correct answer. This lesson has been learned during many moments of great debate in social, political, and scientific history.

Nonetheless here we are. The quest to determine the conditions on Mars before we actually went there is a case example. The uncomfortable lesson, of course, is that science is a human enterprise and, despite the rigorous rules of scientific investigation that are in place, an underlying support for one answer occurred for many scientific debates. With the Age of Reason and the ability to observe and deduce things, we were in a sense just better able to dig a deeper hole to comfortably sit in until further enlightenment in the form of additional observations gave us a useful ladder with which to escape from the hole of our own digging. In the case of Mars, despite the scientific conclusions that Mars was a cold, dry place, there had been a desire that it be like Earth. All of this is a roundabout way of simply saying that Mars has been at the center of a fascination with the potential for "another Earth" or with life beyond Earth since the very first time anyone entertained the notion that it is another planet with a geography and history of its own, going back to the Enlightenment.

MARS GETS PERSONAL

I FIRST CAME INTO THE MARS scene at the very tail end of the telescopic era in the fifties and early sixties just before the space age, a time with all its speculation and excitement about actually going and looking at the planets instead of peering at them from afar. Perhaps it was a result of all the science fiction movies in the fifties, but by first grade, I was an avid watcher of the skies. I recall seeing the first satellites crossing the night sky and thinking that the exploration of space was a thing that was coming into being. I wanted to be a part of that. But I was not particularly fascinated with the aeronautics of space exploration; rather it was the chance to see a new alien world like those in the science fiction movies.

As the space age ramped up, even the US Air Force was involved in preparations for exploration beyond Earth. As part of the development of the knowledge base for eventually going to Mars, a map was prepared of what we knew about Mars. This was the famous Air Force map of Mars, which was essentially the Antoniadi-Schiaparelli map toned down a bit. Hints of the *canali* were still there, but it was the best thing available in the late fifties.

This map, with its faint tracing of lines and blotches on a dull reddish-pink surface, was the standard for all illustrations of Mars in fiction, speculation, and drawings presented in popular astronomy books through the 1950s and even up until the beginning of the 1960s. Mars looked so enticing and alien.

Throughout the entire telescopic era of Mars exploration, the surface details were always just at or beyond the limit of visibility. It did not help that given the limited resolution of Mars with even the biggest telescopes, all we could see were shades and spots that, as we shall see, were completely unrelated to the actual physiographic surface. Mars was really playing a game with us. This is the dangerous zone of human perception: it is easy for the brain to fill in the details where the eye is straining to see something. We were working with limited information. We were trying to solve the equation that was Mars, but we were missing the numbers for many of the variables in that equation. And because we were looking at surficial shades and spots, we were not even using the right equation.

But all that was about to change. We were finally entering the space age, sending spacecraft out to Mars, and we would actually see what we could see. On the eve of the first spacecraft flights to Mars, in the early 1960s, the excitement was palpable, almost like waiting for the next installment of a popular drama series, or another book in the John Carter of Mars series. The stage was set to expect something that we probably had not seen before. We were going to start exploring Mars for real.

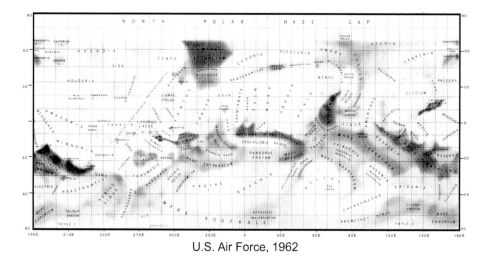

U.S. Air Force, 1962

THREE

The Invasion of Mars, Phase 1

"Nothing spoils a good story like the arrival of an eyewitness."
—MARK TWAIN

FIRST UP-CLOSE EXPLORATION: DEAD OR ALIVE?

WAS THERE EVER LIFE ON MARS? Is there life on Mars now? It's a common but frustrating question. Like every other stage of our collective study of the red planet, we have been jerked first this way and then the other way with each new observation and have been for the past one hundred years of telescopic observations. These alternating highs and lows in hopes for Mars continued when we first became technologically capable of sending spacecraft to Mars during the initial days of modern rocketry and space exploration.

All the publicity and popular culture about Mars had fostered speculation at the outset by early rocketry visionaries like Konstantin Tsiolkovsky and Robert Goddard about the potential for use of rockets to do planetary exploration and to make trips to Mars. In the late 1940s and early 1950s, the famed rocket scientist Wernher von Braun pursued the technical concepts of rocketry and engineering necessary for a Mars expedition in his nonfiction book *The Mars Project,* which became the basis for a series of publications, art, and documentaries that was popular in the 1950s. These are known to have inspired many of the early pioneers of the emerging American and Soviet space programs. It certainly inspired many children of the space age, including myself.

Mars was beginning to figure prominently in Earth's aspirations for space exploration. To paraphrase H. G. Wells's opening in his classic *The War of the Worlds*, but in reverse: across the gulf of space, minds regarded Mars with questioning eyes, and slowly and surely drew their plans. We began enacting those plans by throwing spacecraft at Mars almost as soon as we learned how to launch them out of the atmosphere. The 1960s saw a veritable armada of spacecraft sent to Mars, and this is when Mars began the slow process of resolving from a disk in a telescope to a physical world. Slowly but surely, the Earth invasion of Mars had begun.

The Soviet Union was the first out of the gate in the race to explore Mars with launches not long after the first successful Earth-orbiting satellites. In late 1962, they prepared a launch of a relatively basic probe to Mars, **Mars 1**, in an effort to be the first to send probes to another planet. Mars 1 experienced a variety of problems on its cruise to Mars, culminating in a failure to communicate before arrival. While it did arrive in the vicinity of Mars, without the ability to communicate with Earth, the flyby occurred and the spacecraft continued onward into an orbit about the Sun.

In 1965 the US probe **Mariner 4** was the first to successfully fly by Mars at close range and image a small strip of the surface, returning a series of twenty-four fuzzy—but the first—images of the Martian surface.

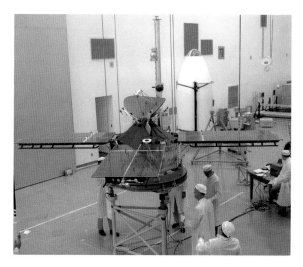

ABOVE: The Mariner 4 spacecraft during assembly at Jet Propulsion Laboratory (JPL). Mariner 4 was the first successful spacecraft to fly by and image the surface of Mars.

It was a shocker. This was the first attempt to acquire direct, close-up evidence of what Mars is really like. I was already an avid amateur astronomer, and more recently I had adopted the lunar and planetary space missions of NASA as a cause worthy of my absolute attention as only a young science geek could.

It was on July 15, 1965, during summer vacation from middle school, that I eagerly awaited the first flyby of Mars by the interplanetary spacecraft Mariner 4. I had read about the Mariner 4 mission long before it launched, and I knew precisely when it would pass by Mars and take the first images of the red planet from close up. To me, it was so obviously a pivotal moment in human history that I expected there would be a blow-by-blow account of the mission milestones as it reported back by signals the success of its interplanetary photographic assignment.

With those thoughts in mind, I waited by the radio for news for the next twelve hours after the roughly 8:00 a.m. EST Earth time of the closest approach. Surely, like the human spaceflight events, I thought, there would be a tremendous hoopla about this historic interplanetary event. But instead there was nothing. No news stories whatsoever of the event. In those days, of course, NASA was not accustomed to the publicity that many space-exploration events are expected to have now. There was no immediate release of mission status or reports of early results. At least none that I could find twisting the radio dial back and forth. I went to bed wondering if I would ever know what happened. It was obvious that I had no clue what the general media thought about important space-science events and that for all I knew there would never be a public statement unless there was something exciting to report. Maybe the images would be revealed someday when somebody got around to sending them to the media and that outlet had a slow news day.

I was surprised that the next morning there was a very short front-page newspaper article stating that the Mariner 4 Mars flyby had occurred and that images were taken of the surface. But, in all the lead-up to the mission, the actual time scale for receiving the data was on the editorial chopping block for most discussions, because, as it so happened, the communication with Mariner 4 did not permit return of all twenty-one images very quickly. In fact, those images were captured on video recorders inside the spacecraft, and these had to be played back over a period of days. But eventually they started coming back, and lo and behold, the newspapers even published a couple of the small, grainy rasterized images! They were a shocker at the time.

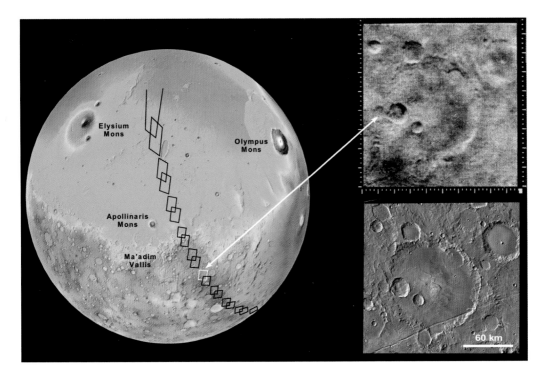

ABOVE: Footprints of all of the images acquired by Mariner 4 as it flew by Mars on July 15, 1965. **TOP RIGHT:** One of the clearest of the images acquired by Mariner 4, the 151-kilometer-diameter crater Mariner, from 12,600 kilometers above the surface. The image is 250 kilometers by 254 kilometers across and centered at 65ºS, 196ºE. **BOTTOM RIGHT:** A much higher-resolution version of the same scene acquired by the THEMIS camera of the later mission Mars Odyssey. The fracture on the bottom is Sirenum Fossae.

Until then, we had developed a fantastic mythical Mars that must be dynamic and somewhat Earthlike with an atmosphere and polar caps and seasons. The surface was surely marked with strange new landscapes of an alien geologically active planet. Instead of fabulous landscapes of river valleys and Earthlike geology, the images showed the surface of a planet with multiple overlapping big-impact craters.

Mars looked just like the Moon! No evidence of life or even dynamic Earthlike processes of wind or water. More to the point, the operative phrase—and headlines—in the press and in the science circles was simply "Mars Is Dead." These details of our new view actually continued to emerge over the coming months because this was the era long before the internet and the availability of instant information for the engaged enthusiast. Most of the images would not be seen by many of us despite our interest until the next issues of our favorite astronomy magazines started showing up in our mailboxes. My source of space exploration information was *Sky & Telescope*, as no doubt it was for many of us at the time. The pictures published were only a slight improvement over the one or two pictures published in the newspapers.

BUT WE WERE NOT TO GIVE UP so easily. Let the press shrug their shoulders and declare Mars another barren Moon and of little more than academic interest. Fortunately NASA had embarked on a program of planetary exploration that included Mars missions at each favorable launch opportunity. So we were still likely to get another crack at exploring Mars with even more-capable spacecraft in a couple of years. Because of the relative motion with Earth in its slightly longer period of orbit, Mars has its closest approach to Earth about every two years. In fact two more Mars spacecraft, **Mariner 6** and **Mariner 7**, were sent for flybys of Mars in July 1969. Interestingly enough, the encounter occurred just a few days after the landing of Apollo 11. In the hoopla around Apollo 11, of course the excitement of yet another flyby of Mars, the "dead" planet, was somewhat muted. I confess that my attention was muted because Mariner 4 had derailed the locomotive of Mars progress for me. Mars was still an interesting place, but craters could only mean one thing: an old dead place like the Moon. Little did I know that "old and dead" and "interesting" were not to be confused as the same thing. Nor was "dead" a particularly good summary of Mars.

I was in the last stages of packing for my first year of college at North Carolina State University, and I was otherwise distracted. Besides, reading about the flybys was not the same as actually participating in the exploration. My college plan was being formulated to allow me to become one of those scientists who got to see all these things for the first time. But there was a problem with planning to be a Mars scientist. What did one actually study in college to be a Mars scientist? There was no precedent for a career track that did that, and there was no university research infrastructure for that topic at the time. Traditionally it would be astronomy, but the study of planetary surfaces was not exactly the same as planetary astronomy. I already had a good background in astronomy from my early years of backyard astronomy, and it was a subject that I had read a lot about growing up. Besides, the science at that time was dominated by all the stories of Mars observations over the last few decades by famous observers, including strange stories of markings on the surface that changed and efforts to determine fundamental things about Mars such as the composition and density of the atmosphere. It seemed like some sort of astronomy was the obvious route for becoming a "Mars scientist." But it was not quite the right area either since spacecraft were not the same as telescopes. And the surfaces of Mars or the Moon, for that matter, were not just objects of telescopic investigation in which one used all sorts of spectroscopic information to determine global characteristics like atmosphere and surface temperature.

Until the problem of career specialization was resolved, my initial plan was to go with some sort of astronomy-physics-type curriculum. But frankly, that did not seem very interesting, and it proved to be rather dull from my perspective during the first months of undergraduate college courses. It took a while to determine why it all appeared off the mark.

What was the thing that made Mars or the Moon, such exciting places? It took a semester or so, but I finally arrived at the explanation with some soul-searching introspection regarding why I was really interested in the study of other worlds. After all, in all those science fiction movies of the fifties, people landed on desert planets that looked, interestingly enough, like early guesses at what a place like Mars might look like, and had adventures while exploring all sorts of craggy geological landscapes. It was the landscape. And the landscape is basically geology. It was obvious. Geology is the surface of a planet like Mars. The more you knew about geology, the more likely you would be in a good position to engage in the early exploration of Mars. I tilted full steam into geology and eventually learned to "read the rocks," as geologists say, and further still I found that as a field geologist, I could map them. This was to prove a momentous course, because I had embarked on a career that had one foot on Earth and the other foot on other planets.

One problem with being in a science path at college was the intensity of the study and almost complete disconnection from the world around me. As a college

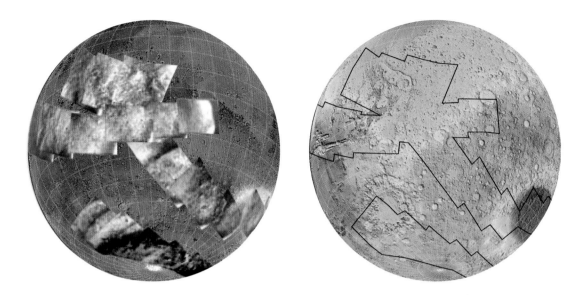

ABOVE: Mariner 6 and 7 images across Mars during a flyby in 1969. Once again a combination of factors obscured details in the more interesting areas such as the vast canyons while the cratered highlands were more clearly imaged.

student deeply immersed in the details of curriculum and, more specifically, making good grades, I was really in no position to watch miscellaneous TV programs or read newspapers. World events, including space missions, were out of sight and out of mind in this environment. I might dip back into the real world when I went home for a holiday, but other than that it was work, work, work. And so, it came to be that another Mars flyby was an event that I may or may not even hear about for a while.

In any case, there was no earth-shattering news coming from missions to rip the attention of a freshman from the textbooks. So, Mariner 6 and 7 happened in 1969, and the event went unheralded in any way that might have broken through the academic fog. Mariner 6 acquired seventy-five images, of which twenty-five were acquired during the close approach of 3,411 kilometers of Mars on July 31; on August 5, Mariner 7 flew by at a distance of 3,408 kilometers, acquiring thirty-three of its total ninety-three images at close range. Another instrument definitely determined that the atmosphere pressure was six or seven millibars, or one-hundredth that of Earth, which was in line with where estimates were going for a long time. But this was a real measurement.

In another historically bad bit of luck, once again the Mariner 6 and 7 image swaths with the clearest views across the surface were mostly within the southern hemisphere that we now know is the heavily cratered half of Mars. In a twist of fate, the images did cover some of what would later be known as some of the more exciting areas, such as the great canyons, but on this day there were obscuring elements, either clouds or dust, in that region that made it impossible to see those features. Mars just was not ready to reveal that little secret yet and, like Coyote, wanted us to do a little more work before we could discover that interesting characteristic.

The Mariner 6 and 7 images revealed more craters and there were no images of anything dramatic like channels, valleys, or even volcanoes that we later learned were present on Mars. Actually there were some published papers that attempted to show that there were some channels and similar features of interest, including one by the geologist who would become my academic advisor in graduate school, Wolfgang Elston, but few paid much attention to those because the images were still a bit fuzzy and such ideas were just geo-guessing. From this second set of images, to the first order of examination at least, it was looking pretty much like Mars was going to be a bust after all, and the hope and aspirations of scientists and of fiction writers were being nailed into a coffin of disappointment along with other forgotten and discarded human aspirations. Nor did it help that the Mariner 6 and 7 encounters took place a few days following the successful first landing of humans on the Moon. A dead Mars was old news; Apollo 11 and the Moon program, now that was the thing.

This was the view of Mars in the early days of spacecraft exploration and would stay that way for two more years. The sixties were not kind to Mars exploration. The seventies were when things started looking brighter for the existence of a more lively and exciting Mars.

MARS: IT'S ALIVE! IT'S ALIVE!

FORTUNATELY, SCIENCE IS A VERY CAUTIOUS process and we tend to hammer a problem until we are very certain that we have the answer. Scientists don't jump to conclusions based on first results, perhaps not because of a particularly righteous scientific discipline so much as we were beginning to get an odd sensation that Mars was going to be a particularly tough nut to crack. A few crummy images that were difficult for the eye to make sense of were not enough to give the answer. Yes, there were craters. But more important, there were a heck of a lot of other things going on in the images that made no sense and could be something new that we were

RIGHT: In 1971 Mariner 9 was the first spacecraft to successfully go into orbit about Mars, imaging over 85 percent of the surface. It revolutionized our understanding of Mars, revealing it to be far more exciting than the results of previous Mars flyby missions had indicated.

unprepared to identify. And we had only gotten images of a very small strip across Mars. So we went knocking on the door of Mars again. In the next Mars mission opportunity, two years after the Mariner 6 and 7 missions, NASA launched **Mariner 8** and **Mariner 9** and the Soviet Union launched the **Mars 2** and **Mars 3** probes, which would arrive about two weeks after the Mariner missions. The two probes sent by the Soviet Union, Mars 2 and 3, were to go into orbit and release soft landers. The Mariner 8 and 9 spacecraft were going to spend some time at Mars by going into orbit and getting a better global view of the place. Mariner 8 failed to achieve orbit during launch and is now somewhere in the Atlantic Ocean. Mariner 9 prevailed, successfully departed from Earth, and went into orbit around Mars in November 1971.

Unfortunately, all three spacecraft, Mariner 9 and Mars 2 and 3, arrived right in the middle of a Mars global dust storm. It is difficult to express the frustration after finally arriving with a decent camera at relatively low orbital altitude looking down on this mysterious planet. At last, we were in a position to get the view we had been

RIGHT: This partial image was returned by the Soviet Mars 3 following its descent and apparent landing on the surface of Mars on December 2, 1971, at 45°S, 202°E, in Ptolemaeus Crater. The image consists of about seventy scan lines and either shows the Martian horizon or is just radio static.

BELOW: Possible images of the Mars 3 lander on Mars acquired by the High-Resolution Imaging Science Experiment (HiRISE) camera on NASA's Mars Reconnaissance Orbiter in 2007.

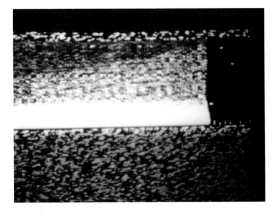

Candidate Hardware for Soviet Mars-3 lander
Landed on December 2, 1971

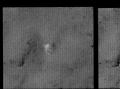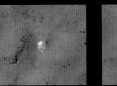

Candidate parachute

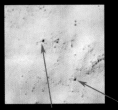

Retrorocket and lander candidates

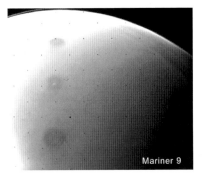

Mariner 9

MGS MOLA
Relief Map

LEFT: The "discovery image" by Mariner 9 during its first orbits of the red planet showing three enormous volcanoes previously missed by flyby spacecraft. The volcanoes were sticking up through a global dust storm at the time. **RIGHT:** The same scene simulated with relief data from the Mars Global Surveyor Mars Orbiter Laser Altimeter instrument from the late nineties.

wanting. But no! Mars was playing games with us just like we had suspected it was doing earlier. At every step of Mars exploration there has been some type of drama, and this was to be the case with Mariner 9. In September 1971, two months before Mariner 9 arrived, Earth-based observations revealed that a global dust storm was cranking up. In approach images from Mariner 9 the planet was an opaque bright ball and the surface was even more featureless than in the earlier images taken by previous Mariner flyby missions. Planet-encircling dust events, the current technical term for global dust storms, happen every few years on Mars, and the reasons they start are even now somewhat unclear. Nonetheless the planet-encircling dust events have bedeviled us right up to the present as later landers and rovers have made abundantly obvious. But it was dramatic and frustrating. We had finally arrived at Mars, but there was a global dust storm obscuring everything! You could practically hear a collective "argh!"

Despite the dust storm, the Soviet Mars 2 and Mars 3 landers were preprogrammed to descend to Mars once the two spacecraft arrived at Mars. The Mars 2 lander went silent after it was released into the atmosphere on November 27, 1971. The Mars 3 lander successfully arrived at the surface on December 2 but ceased transmission after the first twenty seconds of transmission of a surface image, which unfortunately was featureless.

Nonetheless, Mars 3 did transmit something, whatever that was, from the surface and was a historic first, if only partially successful, landing on Mars. Later the Mars Reconnaissance Orbiter High-Resolution Imaging Science Experiment (HiRISE) camera acquired images in 2007, showing what appears to be the remains of the Mars 3 lander on the surface. It had made it, but something went awry.

Coyote Mars was about to let us learn something but wanted us to do a little waiting first. As Mariner 9 patiently waited in orbit for months, the dust storm abated, and a curious revelation began. Three dark spots seemingly lined up in a row; the tops of enormous volcanoes were seen sticking up above the global dust cloud.

Later, the three volcanoes in a row would become known as Arsia Mons, Pavonis Mons, and Ascraeus Mons, from south to north. Little did I know at the time that a few years later I would write one of my first research papers about those three volcanoes. One of the volcanoes visible through the opaque dust in the atmosphere would become known as the largest volcano in the solar system, Olympus Mons, a revision of an earlier name, Nix Olympica ("snows of Olympus"), that had been assigned during the early years of telescopic observation in reference to its appearance as a bright spot in telescopes.

The Mariner 9 mission results established Mars as a new world, or another globe to be explored and mapped. . . . Mars had become a globe with its own geography.

The presence of enormous volcanoes, which represent a type of geology completely unlike the cratered surface of the Moon, was a total reversal from the dull, cratered planet that we had come to expect based on the Mariner 4 and Mariner 6 and 7 results up to that point. At first the mind balked because it was unclear why we should see nothing but craters before and now there were these giant volcanoes. Whatever was going on, it was clear that Mars was not dead, at least not as dead as the Moon! According to these new images it was or had been geologically a very active place, active enough to build volcanoes apparently bigger than any on Earth. And as the clouds of dust began to settle out over the next few months things got even wilder.

An enormous canyon system that would extend from east to west coasts of the continental United States, later named Valles Marineris after Mariner 9, was the first of many dramatic features that came into view. And the mission went on to image many other exciting details, such as enormous dry river channels, wind-streaked surfaces with dunes, layered polar deposits, and weather systems. Practically overnight Mars was transformed in our view from a cold, dead place to a place where even life was imagined as a possibility. We had been jerked back to the Mars of earlier imagination. Coyote Mars was having a good laugh at our expense, again. Now you see it; now you don't; wait, now you see it.

The Mariner 9 mission results established Mars as a new world, or another globe to be explored and mapped. Prior to the space age, the planets were disks as seen in a telescope. With spacecraft flying by and now orbiting, Mars had become a globe with its own geography. Mariner 9 not only showed us that geography, but it gave us compelling views from which we could begin the laborious process of working out the basic global geology. Mariner 9 showed us the fact that Mars is divided into two hemispheres, a northern and a southern one with strikingly different characters. The southern hemisphere is a few kilometers higher and pockmarked with enormous craters with peculiar eroded rims. It was part of this hemisphere that the early missions had by unfortunate chance observed during their flybys and had correspondingly left us with the initial impression that Mars was a dead, cratered world like the Moon. The northern hemisphere is much lower in elevation and the home to vast, relatively undistinguished—or so it appeared at the time—plains, and of course many of the bigger volcanoes, enormous outflow channels, and dune fields along with other amazing things like a gigantic canyon longer than the width of North America.

Beyond the early detection of large volcanoes and giant canyons, Mariner 9 gave us the first good look at many of the riverlike channels snaking across the older terrains. Some of these were meandering like Earth's rivers but were generally deep and canyon-like in shape. Also, the arrays of intricate valleys feeding into progressively larger rivers that are familiar on Earth were absent. Instead the heads of many of the channels started abruptly and often at an alcove. The morphology was rather more analogous to canyons on Earth where the headwaters are springlike sources that are undermined and collapse as water flows, a process known as sapping. This suggested some type of groundwater more than simple runoff of precipitation feeding many channels. Others, particularly the large outflow channels that emptied into the lowlands along the transition between the highlands and lowlands, emerged from large chasms. This led to the speculation that these were some sort of fluvial outbursts (floods) from stores of water or ice released as the chasms developed.

Windblown features were clearly a Martian national landform too. Large areas of dune fields crowded around the lowlands near the north polar areas. Many craters had wind tails marking the influence of the crater rims on the airflow and subsequent deposition of windblown materials. This was the geology that was the basis of many later developments in the study of Mars. Mariner 9 had shown us that Mars was indeed an active and exciting world waiting to be better explored. The results only whetted our appetites for more information through more areal coverage and at higher resolution. We had just enough information by this point to recognize a variety of dynamic geologic processes, but there was just not enough information to begin truly understanding how they worked or what it meant in detail for the history of Mars as a planet. Fortunately, the next great mission was in the works, the Viking Project, and that would provide more orbital detail. The big goal of this new mission was to actually land on Mars and see what the surface was really like.

In the interim, there were other attempts by the Soviet Union with Mars 4, 5, 6, and 7. **Mars 4** arrived February 10, 1974, but a computer problem caused the braking engines to fail, and the spacecraft flew past Mars without entering orbit. **Mars 5** arrived on February 12, 1974, successfully entered orbit, but failed after returning 180 images of the surface. The **Mars 6** lander arrived on March 12, 1974. But in an agonizing turn of events, after returning 224 seconds of data during the descent, all contact was lost just before it was to fire its retrorockets. The **Mars 7** lander arrived at Mars on March 9, 1974, but due to a misfire of the braking rocket it failed to enter the atmosphere and sailed past Mars. Coyote Mars was in a particularly mischievous mood that year.

PLANNING FOR THE NEXT GREAT MARS MISSION: VIKING

I WAS IN THE LATER YEARS of my undergraduate geology studies at the time and felt pretty much like a casual, nonparticipating viewer of all this. Today, we have so many options for getting information that is of personal interest. It is difficult to communicate how frustrating it was to get details about an ongoing mission in the early days before the instant availability of mission results offered by the internet. Flyby missions in the late sixties had been not quite as frustrating because they took place as a single newsworthy event. But during the much longer activity of an orbiting spacecraft at Mars, such as Mariner 9, the modern pattern of initial press excitement followed by a diminished level of reporting became the norm and was exasperating. The daily flow of new images and exciting results gradually decreased in the press until there was effectively nothing. You had to wait for the first published reports in *Science* or *Nature* to learn just what was happening, or rather what had happened. And even those reports were short and images were limited, revealing only the most important findings. But the ongoing journey of discovery was just not presented in the media.

And so it was that I continued my undergraduate work and focused on getting to a place where I could participate one day in the Mars exploration process. I recall discussing my desires to do planetary research and the progress of my degree work with my NC State academic advisor, Charles Welby, who taught a great course on Earth's geologic history, also known as historical geology, that necessarily included reviews of the fossil record on Earth. At some point in the discussion he said that this was great, maybe I would be the first person to discover fossils on Mars. Ha! Well, that was a nice sentiment, but I was not particularly interested in fossils. Instead I leaned more toward understanding the geology that produced those wonderful landscapes such as the volcanoes that Mariner 9 was seeing. But who knows what strange twists a career might take?

My first brush with those twists and with actual participation in the drama of Mars exploration happened one day when we had Bill Muehlberger as a special lecturer in my undergraduate geology department. Bill, a vocal advocate of basic field geology based on his years of work in the field, was prominent in the geologic training of the NASA Apollo lunar astronauts. To illustrate how geologic science had become interwoven with planetary science it is necessary to review Bill's involvement in a larger process of the time. Bill was part of a consortium of geologists put together by

Lee Silver of Caltech to train the Apollo astronauts in geology so that they could act as surrogate geologists during their visits to the lunar surface.

The original Apollo program actually had no plans for geologic science on the Moon, since the focus was just getting there and back. Silver realized that an opportunity was going to be lost unless there was geological input to the process, so he had formulated a group that would train the astronauts and help in selecting the most valuable sites to visit on the Moon. That is another involved story best reviewed in the context of lunar exploration. But the importance for planetary science was that as a part of this program there was a search for analog geologic sites here on Earth to help field-train astronauts. To help support this Apollo training activity the US Geological Survey established a field center near the site of the Lowell Observatory in Flagstaff, Arizona, surrounded as it was with analogs like the Barringer Meteor Crater and the young volcanic features of the area such as Sunset Crater. This was to become the US Geological Survey Astrogeology Science Center, which became the home institution for many early pioneers in planetary geology and later the hub for planetary geologic mapping.

Bill Muehlberger was on a tour of universities at the time in order to present some of the results of the just-ending series of Apollo lunar landings. It was during the course of a post-lecture discussion session with students that the subject of the next great NASA missions came up. I do not recall whether I asked Bill or whether the topic came up in general, but a Mars lander mission that was being planned, called Viking, entered the discussion. For the Viking mission there would be two orbiters and two landers. In an aside to his discussion, Bill harkened back to his experience with the whole colorful, technical, and personality-laden process of determining the Apollo landing sites. He said he thought that he could see the clouds gathering for a big battle among the planners regarding where those Viking Landers were going to be set down on Mars. And to make matters more intense, he added, the landers would be orbiting Mars attached to the Viking Orbiters initially while the need for sending the landers to the surface became urgent as the clock continued to tick on who-knows-how-long-or-how-short the mission lifetime might be. And the only reliable way to pick landing sites would be for the Viking Orbiter imaging cameras to acquire high-resolution image maps of possible sites that had been previously tentatively identified from the much lower resolution images provided by Mariner 9. And if those sites turned out to be unacceptable, then the search had to continue posthaste. From

It is difficult to communicate how frustrating it was to get details about an ongoing mission in the early days before the instant availability of mission results offered by the internet.

his experience with the Apollo site selection process that had taken place in a relatively drawn-out fashion to analysis of previous Lunar Orbiter data, Bill imagined the intensity of the real-time debates about the Mars Viking Lander selection process could be epic.

"Boy," he said with a particularly knowing grin, "I am glad I will not be a part of that debate!"

I recall saying something like "Ha! Ha! Me too!" Little did I know at the time that a few years later I would be directly and intimately involved on the front lines of exactly that process!

A PERSONAL JOURNEY TO MARS BEGINS

MY JOURNEY TO MARS SERIOUSLY GOT underway as I was coming to the end of the undergraduate process in 1971 and the need arose to seek a university where I could do graduate school studies, hopefully in a place where actual Mars and other planetary research was taking place. But a funny thing happened on my way to Mars. I discovered that geology was pretty interesting in its own right. In those days there were only a couple of universities where a faculty member was doing planetary research, mostly on one coast or another and not where the exciting geological landscapes were located. But I wanted real field experience that would give me better insight into the geology that we were starting to see on these worlds like the Moon and Mars. If I could not go to Mars, a good option was going to a place that looked like Mars and with many of the same geological characteristics and science problems waiting to be explored and solved. There is an old phrase in geology that says, "The best geologist is a geologist who has seen the most rocks."

So I began my process of searching for a graduate school by looking at universities that were in interesting places. There were some pretty good schools to choose from; not many, but a few. But most of the options were in big cities, or worse still, big cities in green places, not nice rocky, geological places. So for me, the best place meant those places that were mostly out west. One place that caught my eye was New Mexico. At the University of New Mexico, there was a faculty member in the geology department, Wolfgang Elston, who was actually funded by NASA to do some early

> "So one day my mother sat me down and explained that I couldn't
> become an explorer because everything in the world had already been
> discovered. I'd been born in the wrong century, and I felt cheated."
>
> —RANSOM RIGGS, *Miss Peregrine's Home for Peculiar Children*

studies of the Mariner 9 results. Wolf was a volcanologist who worked on the volcanic areas of New Mexico, and planetary geologists had, by that time, realized that volcanism appeared to be a significant part of the geology of the Moon and probably the inner planets. In retrospect, I was very fortunate to work with Wolf. He was soft-spoken and kind and truly cared about his graduate students. He had done foundational work in the interpretation of some of the volcanoes in New Mexico and was well respected in the profession. When you did fieldwork with him he kept up a constant barrage of jokes and funny stories; your smile muscles actually began to hurt after a while. Later I found out that this kind man with all the jokes had survived the Holocaust, fleeing Nazi Germany to spend the rest of World War II in England at a boarding school for refugee children, and made his way to the United States after the war. Wolf was doing research with Mariner 9 data but was also of the opinion that people who wanted to study the new science of planetary geology should really spend time doing field geology so that they had a complete knowledge of how geology actually worked. This business of just analyzing spacecraft data alone was not a good thing in his opinion. I agreed. Even better, I was offered a research assistantship there, which meant I would work with him on research projects and a small stipend would fund living expenses.

THE NEW GEOGRAPHY OF MARS

SINCE MARINER 9 HAD IMAGED MOST of the planet, it was now possible to display the actual surface features of the planet in a global map. This was done in a variety of ways. At one point an actual globe of Mars was created at the detail level of Mariner 9 images and it was done by hand pasting individual images on a large physical globe as a sort of three-dimensional mosaic. In those days, before the widespread use of computer image processing that permits stitching together images without glaringly obvious seams, the resulting map was difficult to comprehend from a distance due to the grid of image margins disrupting the eye's ability to follow the continuity of large features.

One of the masterworks of this process was assembling all the Mariner 9 images into a global map. This is where the US Geological Survey Astrogeology Science Center in Flagstaff came to the rescue. Mapmaking is of course the province of the US Geological Survey, and during the Apollo Moon program the methods for making maps with orbital images were well developed. One of the methods used to get around the problem with the disturbing image edges made from dozens of images was literally artwork using an airbrush to replicate the principal features in a mosaic. Airbrush mapmaking was a specialty and a time-consuming process, but the results were well worth the effort. Now it is possible of course to do the same thing via computer processing and generate a map from many images that appears seamless. But this is how it was done before the widespread utility of computer image processing that enabled the stitching together of digital images.

Besides getting a clearer overall view of the global mosaic for purposes of understanding the global Mars geography, or aerography as it is called, the mosaic could be fitted to a map projection that preserved the details while presenting them in a coordinate grid representing the surface of the particular planet. The center of this coordinate system used the historic precedent set on the Beer-Mädler Mars map of 1832 in which Sinus Meridiani was the prime meridian. But accurate maps require a precision tie point on the surface. Now with the better than one-kilometer resolution provided by Mariner 9, Merton Davies of the RAND Corporation set the meridian, or the 0 point on the longitude grid, of Mars to a small crater in Sinus Meridiani about half the size of Barringer Meteor Crater here on Earth. Mert was a fixture at many early Mars geologic mapping meetings. I recall on several occasions listening to Mert discuss in presentations on the latest updates on efforts to precisely fix the prime meridian of Mars using best fits of all sorts of versions of the Mariner 9 images fitted to the global coordinate grid. It felt like I was sitting in on a meeting of a surveyor's club more than a geological meeting. As a nod to the early work in cartography on Earth the crater used as the tie point was named Airy, after Sir George Airy, who in 1851 had designed the transit instrument used to define the location of the meridian in Greenwich, England. As one may have guessed by this point, Mars exploration history is rich in far-ranging allusions to historical connections. Meanwhile, the whole science of cartography is an exercise in head-exploding precision applied to the fitting of spherical grids to worlds that are fundamentally not quite spherical and the corresponding efforts to manipulate the grid to some sort of consistency.

The most frequently reproduced version of the resulting global map was a Mercator projection, a projection that was used for many planetary missions during the first decade. Later a more plotter-friendly projection of simple cylindrical coordinates became the standard.

On a side note, for various reasons the less geological and more geophysics-oriented scientists have been fixated on maps that presented equal areas, which neither the Mercator nor simple cylindrical projections did with their distortion of high latitudes. The Mercator projection is the global projection on some wall maps of Earth in which Greenland is displayed as a "giant continent." Other projections produce odd elliptical-shaped global maps that are designed to present equal areas but produce mind-bending distortions of actual shapes. In any case, now we had a global map of Mars as seen by an orbiting spacecraft, Mariner 9.

The comparison of the new map made from all these wonderful images of the surface of Mars with the classic maps of Mars made from earlier telescopic observations was exciting and disturbing. What was at once obvious was the lack of correlation between the old Mars of telescopic observation and the new Mars as seen in orbiting spacecraft images. A comparison of the pre-Mariner spacecraft map of Mars, such as the Antoniadi-Schiaparelli map, with the map produced after the Mariner missions shows that the dark markings and white blotches, the features of Mars drawn from telescopic observations for a hundred years, and so much debated during early telescopic studies, had very little correlation with the physical terrain of Mars.

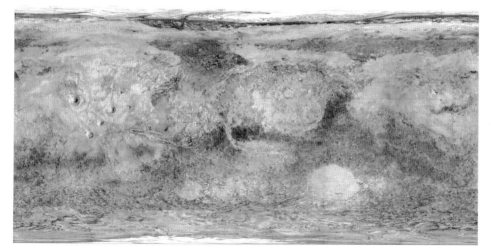

LEFT: Global airbrush map of Mars based on a mosaic of Mariner 9 images. Before the widespread use of digital-image manipulation methods it was common to produce maps using artistic airbrush renditions of features in actual images.

Beer and Mädler, 1830

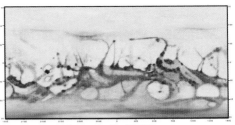

Antoniadi, 1909

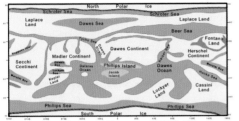

Proctor and Dawes, 1865-67

U.S. Air Force, 1962

Schiaparelli, 1877

Mariner 9, 1971

Lowell, 1901

Viking, 1976

Flamarion and Antoniadi, 1900

MGS/MOLA, 1999

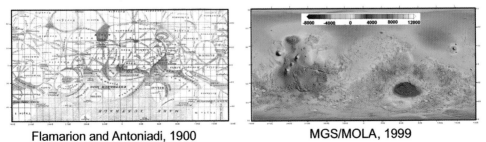

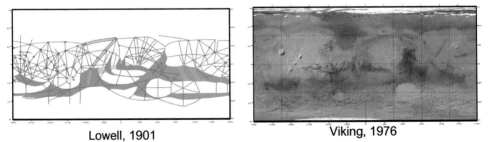

ABOVE: Compilation of Mars global maps. Early maps based on telescopic views of Mars were only able to resolve bright and dark permanent features. Later maps after Mariner 9 showed that the bright and dark features had little to do with the actual surface relief features of Mars. Modern Mars scientists mostly use the color-coded topography relief map from MGS/MOLA.

We now know that the dark areas are really just less dusty areas on the surface where wind has scoured dust or where dust was prevented from collecting by constant cleaning of the surface by wind. These tend to be the more exposed and rockier areas. But the larger-scale relief and the actual terrain or geography of Mars was mostly unrelated to the differences in shading. It appeared that Coyote Mars had played yet another trick on us and had sent us on a wild-goose chase with the large dark markings and bright spots that were so much the subject of telescopic study. Nor was this the last time Coyote Mars sent us off on frustrating quests. More sobering, one wonders to what extent that theme continues with current hotly debated Mars research. This new view of Mars terrain features led to the need to completely revise the system of names used earlier. As we shall see later, many of the old names from the Schiaparelli system of Latin names were kept and merely transmogrified to identify actual surface features in the same locality on the surface of Mars. Others were discarded. So we now have a global map of Mars that bears echoes of those former times. We will revisit the manner in which those names were used a bit later.

Over the course of the years following the creation of this global map, it was further divided into a series of quadrangles that depicted large segments of that global map. Quadrangles, or large four-sided polygonal areas, are the standard method of breaking down a large region for more detailed geologic analysis. These quadrangles were then assigned to various researchers for further study. One of the early projects for Mars geologists that was beginning to emerge was to take those quadrangles and the underlying images, frame by frame, and attempt to draw up geological maps that made some sort of sense of the terrain in terms of relative ages of events that created different terrains. This had been done previously with the Moon and was part of the early days of the new science of planetary geology. The process was based on determining apparent overlapping relationships between terrain "units" via the photographic evidence in a process called photogeology.

On Earth geologists had for many years used air photos of areas that were to be mapped, but only as guides for actual groundwork in the field. But the process was taken to the Moon, where the photogeological maps became an end in themselves. In a few places those maps helped define the questions that were later actually visited by astronauts. The maps provided a template for questions about places that could be addressed with more detail study in the field. But until someone actually visited a place, the photogeologic map was a good first attempt at sorting out what was actually present and helped determine the rough outline of the history of events that had shaped the surface; in other words, the geology. Wolfgang Elston was engaged in the early process of doing a photogeological map of a quadrangle map of Mars, and one of my first projects on arrival was to do photogeologic analysis of some peculiar round hills that populated a broad plain near a mountainous terrain on Mars called the Phlegra Montes.

At around the same time that I had completed that project, I had an opportunity to go to my first scientific meeting. This was an annual meeting of scientists funded through the NASA Planetary Geology and Geophysics Program where everyone gave a brief report of research results on their individual funded projects. I do not recall many details of the meeting other than presenting my research results and listening to other reports from a whole roomful of other Mars scientists. The one thing that I do recall was the fact that there were a lot of scientists there who I had seen in TV interviews throughout the course of the modern space age of spacecraft sent to Mars and on the lunar missions, eminent planetary scientists Hal Masursky and Mike Carr among many others. Wolf had explained to me that going to such a meeting was important for all the usual reasons of networking and such. But he also said that a very important reason was to see for oneself that all these luminaries were just ordinary people, that they were not ten feet tall or had bug eyes or were otherwise in any way gods. They were just ordinary people doing amazing things. It was true, and it was clear that with hard work and good science I too could join their ranks in time. These meetings were a staple for many years and were a major source of insight into new science that was being done and how people were doing it. More important, they were a good metric for seeing how well you were stacking up, or not, as the case might be, with the state of the art.

VIKINGS DISCOVER A NEW WORLD

FROM 1973 TO 1975, WHILE IT was tempting to just dive into the developing world of doing space science research as part of my quest to acquire an actual expertise that would be applicable to the new science, I continued to develop real-world geologic skills by engaging in an Earthbound field geologic mapping thesis. Once again, my attention was divided between the new science of planetary geology and traditional Earth geology, pretty much following the belief that the best geologist was familiar

LEFT: The totally rad author dude in the media briefing room at JPL in 1976 next to a replica of one of the two Viking spacecraft placed into orbit at Mars in 1976. The landers were encased in an aeroshell and heat shield that separated from the orbiters and descended to the surface. On the left is a global mosaic of Mars consisting of Mariner 9 images stitched together and assembled before the Viking mission.

RIGHT: Jayne Aubele standing next to a replica of one of the Viking Landers on display at JPL during the Viking Mission in 1976.

with the established science and that the application of real geologic knowledge to planetary surfaces would follow naturally. It was during the early seventies, while I was seriously involved in just learning how to be a real field geologist, that the next great Mars mission that Bill Muehlberger had mentioned way back a few years earlier, Viking, was well underway. With planning, the eventual launch in August 1975, and the arrival at Mars in June 1976, the Viking mission was going to be spectacular.

Carl Sagan, a faculty member at Cornell University, had become the prominent media face of Mars exploration, and he was doing a lot of public presentations at the time extolling the importance of the mission. It was a full-scale NASA "flagship" mission, and the hype was extensive. The whole launch and eventual landing of the first Viking Lander happened as I was working day and night as a graduate student, taking the required classes, working on research projects, and learning from Wolf how to do this all-encompassing endeavor called science. I also met and started going out with a fellow graduate student in the department, Jayne Aubele. She was a petite blonde, intelligent, determined to be a good scientist, and had a sense of adventure. Our dates were usually Saturday hikes to interesting geological places in New Mexico. And, a few years later, after we had both gone on to PhD work at the University of Arizona, she became my wife and lifelong partner in science.

I was busily mapping in a remote location in New Mexico, following Viking much as I had the Mariner 9 mission years earlier, that is, remotely and loosely but with great interest and desire to be a part of something like it. One day during one of my returns to town in July 1976, as an interlude to the mapping sessions, Jayne showed up at my apartment door. She was also one of Wolf's students, actually his first woman graduate student and one of the few women grad students in geology at the time. She was busy on her own remote geologic mapping project in New Mexico but also wanted to go into the field of planetary geology and Mars. When she knocked on my door, she was holding the newspaper headline with the report of the successful landing of the first lander of the two sent to Mars, **Viking 1**. Prominently splashed on the front page was the first image taken on the surface of Mars by the Viking 1 Lander.

Gosh, that looked very interesting. In fact, the rocks looked pretty much like the ones she and I were mapping in volcanic areas of New Mexico. Every day, at my feet, I saw rounded volcanic rocks like the one next to the Viking 1 Lander footpad. And then, amazingly, as I was about to return to fieldwork, Wolf suggested something that I might want to do. In the course of his NASA work he had met Tim Mutch, a faculty member at Brown University who was the principal investigator for the Viking Landers that were just then arriving at Mars. Tim went on to write one of the first Mars textbooks based on his research with the early Mars missions through the Viking results. Now the Viking Project had created this little program called Viking Student Interns. The idea was that these students would go the center of Viking Mission Operations at the Jet Propulsion Laboratory (JPL) in Pasadena, California, and help with tasks in the mission operations. The program was already underway, they had interns, and the program was for undergraduate students. But Wolf asked Tim if he could send one or two graduate students out. Tim said sure, but the funds for the program were already assigned, so Wolf would need to pay for my participation out of his Mars grant funds. So even though I was a graduate student, not an undergraduate, I would be able to go to the center of Viking Mission activity at JPL as a kind of free agent and with considerably more geological background than an undergraduate. Besides, the geology department already had a faculty member, Klaus Keil, head of the Institute of Meteoritics, who was on the Viking Lander team in the capacity of a geochemical expert and member of the lander group operating the chemical analysis device. Although my interests were more broadly geologic rather than chemical, there was that connection.

With that, in late August 1976, just a few weeks after the first successful landing on Mars by the Viking 1 Lander, I packed up the car and drove from Albuquerque out to JPL into a sunny but much lower elevation, humid, and very smoggy Pasadena, California. I had just spent the previous months stomping around at nine thousand to ten thousand feet elevation during my field mapping, so the low elevation felt like I was in a pressure chamber and decidedly unlike the thin atmosphere of Mars. As an aside, I should note that 1976 was just before the smog problem had begun to be addressed seriously, so the haze of blue fog was fairly thick. In fact, it was so thick that

> Compared to the Viking Lander team workspace the space where the Viking orbiter imaging team was camped out was open, brightly lit, and relatively bustling with activity.

I recall that a week had passed before I realized that there were mountains right next to Pasadena. Nonetheless in this smoggy, new, low elevation world I found my way to JPL, got all the badges and clearances for access to the mission operations, and eventually arrived at the holy of holies, the Viking Lander team mission room in Building 264.

It was a rather dark room, and it appeared relatively quiet, as things had settled down after the landing, initial panoramas of the landscape had been done, and general monitoring and soil-sampling activities were underway. I found Tim Mutch in his office in the corner and reported for duty. Tim was a very academic-appearing, tall, and somewhat lanky fellow with round glasses and a professorial bearing mixed with a bit of the hard and direct administrator demeanor. When I caught up with him, he was busily working at his desk in the peace and quiet of the lander team workroom, probably a manuscript on the first results from the Viking 1 Lander. During the course of introductions he said that things were kind of slow with the lander now, as was obvious from the deserted character of the room, and the main activity, and where extra help was needed, was probably over in the Viking Orbiter team.

The second lander was still attached to the **Viking 2** Orbiter and there was a big push to find a landing site and get that lander down on the ground. He felt that the Viking Orbiter team could really use some help. Well, gosh, everybody wanted to be on the lander end of things, since that was where the really new thing was happening, or so it seemed. But I certainly had a sense that given all that interest by people with more experience than me, the opportunity to contribute something did seem a bit daunting. And the quietness of the room suggested to me that the hoopla was definitely waning on the lander side of the story. So, I was not terribly disappointed when Tim walked me over to the next room to talk to Mike Carr, head of the camera team for both Viking Orbiters. Compared to the Viking Lander team workspace the space where the Viking Orbiter imaging team was camped out was open, brightly lit, and relatively bustling with activity. All sorts of images consisting of mosaics of many images depicting the candidate landing sites were arranged along the walls, and stacks of new images and binders with recent images were lying about on the team conference table. Mike was in his office looking harried and preparing for a daily landing site selection committee meeting that took place every afternoon, but he stopped long enough to make introductions.

Mike Carr was an interesting fellow, and of course one of the greats of Mars science. He was intense, English-accented, with a compact build, and spoke like he was actually thinking about something else but was giving some other half of his mind to the conversation at hand. He also turned out to be kind, considerate, and an excellent planetary geologist. Once introductions were made, he thought for a second and said that I could probably help some of the Viking Orbiter camera team members in their ongoing work analyzing the images being returned of the potential landing sites for the Viking 2 Lander.

Several potential landing sites were on the table. The one thing they had in common with the Viking 1 Lander site was that they were in relatively low elevations. One of the important goals of the Viking Landers was the search for possible life or evidence of past life, and there were specific instruments on the landers to search for the chemical traces. The candidate landing sites were chosen to maximize what was thought to be the best chances to look for life. From Mariner 9 we knew that there were vast outflow channels and dry riverbed-type features that indicated water had at one time flowed on the surface. Some of these channels emptied into basins where it was thought that either water or water-saturated sediments might have been present at the time the channels were formed. Because water was unstable at the low atmospheric pressure of Mars, it seemed that the best chances of studying anything having to do with water, like past life, would be in the lower elevations. Besides the science reason for looking at low elevations, there was the fact that any spacecraft that would need to use the atmosphere to slow down and especially to use parachutes in the final stages of descent would need a bit more atmosphere to do so than the thinner atmosphere of higher elevations. So, the course was set very early to look for the kinds of data relevant to past life in lower elevations and especially at the mouths of large outflow channels. The orbiter imaging team consisted of a wide array of the Mars luminaries at the time. This included some people I was familiar with from previous meetings in the years leading up to the Viking mission, such as Hal Masursky, who had trained the Apollo astronauts, and planetary geologists Ron Greeley, Arizona State University; Jim Cutts, JPL; John Guest, Open University, London; and Larry Soderblom, US Geological Survey.

ABOVE: Building 264 at JPL on an unusually smog-free day in Pasadena, 1976. This is where the Viking mission operations took place. In the foreground is the mall in the center of the JPL campus.

RIGHT: Photo of the Viking Orbiter image team in the orbiter science team area of Building 264 in September 1976.

When Mike walked me into the orbiter team work area it was mostly empty except for John Guest, who was sitting quietly at a light table working at photogeologic mapping on a recently assembled Viking Orbiter image mosaic of one of the favored candidate landing sites for the Viking 2 Lander as part of the ongoing characterization for site selection discussions. John Guest was the American stereotype of an Englishman, professionally attired in coat and tie and dark horn-rimmed glasses, and with a soft voice and an accent that sounded like *Masterpiece Theatre*. His manner was one of great politeness, and when he spoke to you it was with a familiar and enthusiastic delivery as though you were the most important person he had occasion to speak with that afternoon, perhaps over tea and biscuits.

After introductions, Mike Carr returned to his office and John proceeded to explain to me what he was doing. He had a sheet of Mylar overlain on the mosaic and was carefully tracing out interpreted geologic features, drawing lines—which we geologists call contacts—between obviously different terrains or geologic units. In those early pre–personal computer days, everything was of course done with hard copy, so large-format mosaics were the way we interacted with the mission images. The mosaic was of an area in a plain known as Utopia Planitia, actually just to the west of the Phlegra Montes that I had been researching a year or two earlier. It was selected as a possible landing site because it was relatively flat without much more than a few craters of various sizes, and, more important, it was a low-elevation area considerably below the mean Martian surface elevation. Unfortunately it was a mess of small one-kilometer-size little hills, or so it appeared. Because the contrast had been "stretched" (enhanced to show very bright and very dark contrasts), every minute detail was made to appear like something more important, so the hills were actually little more than gentle swells that one would be hard-pressed to recognize even if you were standing on them.

But the really unfortunate thing was the resolution of even these new Viking Orbiter images was too low to really resolve the surface at the scale of the lander. This meant that it was not possible to say with certainty that a given point on the map was free of things, like large rocks, that would damage a lander with low ground clearance. So the object of the mapping exercise was to determine as much about the apparent geologic process as possible, and using that understanding one could infer something, anything, about the likely surface characteristics at the lander scale.

Examination of the surface using the images acquired by the Viking Orbiters was important for the process of selecting a safe landing site. Every piece of information that could provide some insight into the nature of the surface at the scale of the lander, or about two or three meters, was being thrown at the problem.

Examination of the surface using the images acquired by the Viking Orbiters was important for the process of selecting a safe landing site. Every piece of information that could provide some insight into the nature of the surface at the scale of the lander, or about two or three meters, was being thrown at the problem. And new data were arriving daily as the Viking Orbiters continued to acquire new views of the surface, with particular focus being given to the candidate landing sites. In the weeks leading up to my arrival on the scene, three sites, a primary, secondary, and tertiary site, were being considered for the number 2 lander. And they were all looking less than desirable in terms of apparent complexity and the potential for lander-killing surfaces. Meanwhile remote sensing studies were being done using Earth-based radar and with infrared. Hugh Kieffer was the Viking Orbiter team member who used another Viking instrument, the Viking Orbiter Infrared Thermal Mapper (IRTM), to assess the physical character of the surface. He was the guru of using infrared wavelengths of light to get at the heart of the question of surface roughness by hammering away at the fact that you could use the thermal properties revealed in the infrared to estimate the size and abundance of rocks on the surface. At least it was a way to go beyond the spatial resolution limits of the visible light–imaging system and say something quantitative about the surface roughness at lander scales. Because these new data were relevant to the very real and immediate quest to get the second lander on the ground pronto, there were daily meetings of the landing site selection and verification committee in which updates on all the methods being brought to bear on the problem and the results were presented.

The legendary tough boss of this process and lead at these meetings, the person who was responsible for making successful use of the taxpayers' dollars, was the JPL Viking Project manager Jim Martin. Jim was a big, serious engineer with a crew cut, and when he spoke folks listened, obeyed, and maybe even felt as though they wanted to snap to attention! What he wanted first and foremost from all this scientific poking and prodding at the problem of site verification were facts and something closer to certainties. Efforts to analyze the candidate sites needed to show some real results, not speculations and academic theories. And those efforts were needed now, not in some publication with theoretical positions stated, tested, and debated.

This was an engineering problem in which we needed to know something, not just provide our best guess. Guessing is not an accepted method in engineering. The engineering method is just like the scientific method but more so in many ways, requiring careful documentation of parameters that are then used to make calculated predictions. Do we know with certainty that the bridge is going to hold up, or is this just a guess and it may well collapse at any minute?

In a high-powered meeting with limited time for presentation, the prize goes to those who can present their results in the most compact manner.

Do we know with certainty that there are rocks of such and such a size on the surface, or is this an educated guess? What can we determine that we can then use to make serious estimates? This clash of cultures between science and engineering has often been highlighted during planetary missions because mission science teams work cheek by jowl with mission engineers. But it is a good kind of clash, and I think that better science is a result. We scientists tend to drift from the scientific method in the atmosphere of academic arguments but get reminded of the purer aspects of the scientific method through the interaction with engineers. I know that I certainly have benefitted by the interaction repeatedly.

This is where the efforts on the Viking Orbiter imaging team, such as those of John Guest and others, were relevant. We needed to show some serious results that resolved the problems associated with deciding where to land safely. The first two sites were starting to look like they were going to be a hard sell. The terrain was too rough and too complex to know if these were going to be safe places. So analysis of the third site in Utopia Planitia was needed. I do not recall whether it was John or someone else, maybe Jim Cutts, who suggested that what I could do was sit down with the images of the Utopia Planitia mosaic and make some kind of estimate of visible "hazards" and build a map showing where the surface was unfavorable and where the surface looked maybe okay.

So there I was, doing precisely what I had laughed with Bill Muehlberger about hoping to avoid a few years earlier. It was obvious that the tangled web of fate was played like a violin by Coyote Mars, and in the farthest corner of its influences. I sat down and scanned the images, making up a list of all the different types of features and ranking them according to how bad they were. There were craters, and those were bad, very bad. Craters were deep holes, maybe with rocky or dust-filled bottoms. Worse yet, craters frequently had rocky rims, and worse than that, they had steep slopes. Craters were all kinds of bad. The presence of a crater right down to the limits of Viking Orbiter image resolution, which was by today's standards an absurdly large one hundred meters or so, was not something you wanted to put the Lander down on top of. Then there were all these hills and bumps. Same thing there; slopes were bad, although it turned out later that the slopes were hardly noticeable on the ground but appeared notable only because of the high contrast in the mosaic. But in the heat of the landing site search, uncertainty was not a good basis for selecting an area. Hills were to be avoided. Then there were large areas that appeared dark and speckled like they were rocky or something. Others thought that they were dunes. What that something was we could not say, but they looked like badness. Cross those dark and speckled areas out.

Building this map and identifying all these bad things to avoid on landing would take time. During this process, it was suggested that I attend the daily landing site selection and verification meeting at around four o'clock every afternoon to see what we were up against and get a sense of what I was going to need to do to support the team. The meetings were held upstairs, in a small room with tables arranged in a U. At the front of the room there was a screen and a viewgraph, or overhead, projector. After I arrived in the meeting room, a group of what seemed to me to be very somber people began gathering around the table. The "principals," that is, the people who were formal members of the committee and who presented and voted on things that needed to be decided, sat at the table. The rest of the "second-string players" and us "hangers-on" sat in chairs behind them against the walls. Jim Martin started each meeting reading the riot act, or so it seemed to me, for whatever the topic of that meeting was going to be. Then, one by one, the various representatives of the differing efforts to analyze the candidate sites got up and presented the latest results via a stack of eight-by-eleven transparencies that they each hand-placed on the viewgraph projector.

Over the course of a week or so of these meetings, people like Hal Masursky, Mike Carr, Tom Young, Hugh Kieffer (folks who would be later recognized as among the founders of the field of planetary geology), and others got up and delivered quick, deliberate presentations about the results of their team's assessment of the sites over the last day using their individual methods and the latest data to come down from the orbiters. In most of the meetings that I attended, Hugh Kieffer was getting a lot of traction with his thermal inertia rock abundance estimates using the Viking Orbiter IRTM instrument. Thermal inertia was an interesting new take on the process of estimating rock abundances. The way it worked was to use the fact that big rocks tended to stay warmer than small rocks and sand grains. The bigger the rock, the longer it takes to cool down at night. Think of beach sand. In the full sunlight of day, beach sand can be so hot it can be painful to walk on. But after sunset the same beach sand is pleasant, maybe even cool. Meanwhile any large rocks on the beach are warm far after sunset. On Mars, a similar effect is present. The more rocks, the greater the temperature relative to the surroundings. So by doing a survey of a surface with an instrument that viewed the surface in the infrared part of the spectrum it was possible to say where rocks were more prevalent.

But it was a new science and there were many variables that were still being worked out. And the work with radar estimates of roughness from Earth-based radar at Goldstone were not far behind in providing some degree of numerical certainty in an otherwise fuzzy world of trying to make sense of safe areas to land. Radar signals from Earth were sent out to Mars and the returned signal would be scattered where the surface was rough and not so much where it was smoother. The actual technology is somewhat more sophisticated, but that was the general principle, and it had had some success in early efforts to characterize the surface of the Moon during Apollo landing site selection and verification efforts. The radar results were an area of new science and somewhat formative, but nonetheless provided some sort of numbers based on actual observational measurements. By contrast the imaging team's arguments appeared more like arm waving given the scale of the features that the images could really say anything about. We image team types needed something more quantitative to bring to the site selection table.

One day as we labored on the analysis of the images for landing site selection, Carl Sagan, a member of the Viking Lander science team, wandered through the Viking Orbiter science team area. He was about to do some big press conference and was looking over our shoulders specifically asking about anything new and interesting that was going on with the orbiter imaging. I was taken aback. Here was one of the gods of Mars science, and he did not have absolute knowledge of what was already imaged. Then it dawned on me: maybe it was not so bad being a mere foot soldier in the great quest for Mars science. At least we foot soldiers got to look at the images every day. After that I realized that sometimes just being a common Mars scientist might not be so bad. We got to see things and see details that the gods did not see.

There were several sites being considered, and for each there was a need to select a point for a landing ellipse. Landing ellipses are imaginary areas surrounding a target point in which the ellipse represents a zone over which there is a high probability of actually landing. The reason that the area is elliptical relates to many uncertainties in both knowing where Mars is relative to the spacecraft and knowing how the lander interacts with the atmosphere on landing. So if there is a point that you select, your probability of hitting that point is spread out but there is a 100 percent probability of landing within the ellipse and a somewhat smaller but significant probability of landing within the central part of the ellipse. As a reference we also used a smaller ellipse inside that bigger one that was about half the length and width and represented the 50 percent probability cases. Landing ellipses have been getting smaller and smaller with experience at landing on Mars, and are down to a ten-kilometer circle these days, but in those days the best that we could do was an ellipse that was about 100 kilometers wide and 250 kilometers long.

Ultimately the desire in this case was to find an area on Mars on which you could fit an ellipse with the minimum hazards to landing. And of course the underlying desire in all of this was to find a place that was actually of interest for the mission goals. It was all about coming to an agreement of where the very best landing ellipse could be located on Mars. And that decision really needed to happen very soon.

There were several of these meetings leading up to the time for a final decision. In the meantime, I went back to work on making a presentable map of the landing hazards that the team might find useful in selecting a site for the landing ellipse. The idea arose to make a contour map of the level of hazards. Since I already had a map of the Utopia Planitia candidate landing site with small numbers labeling each of the hazard features according to a scale of undesirability, it was a simple matter to overlay a one-kilometer grid and sum the identified hazards in each grid box. Boxes with higher sums were bad. Boxes with lower would be better. Then using this grid of numbers that estimated the level of hazards it was a simple matter to draw contours that followed the numbers representing the level of hazards per square kilometer. The reason for presenting the results with contours was simply that as a graphic device they permitted the variable degrees of badness and goodness to be more easily visualized and digested at a glance.

In a high-powered meeting with limited time for presentation, the prize goes to those who can present their results in the most compact manner. In those days, it was not uncommon to do manual contouring by interpolating by eye between numbers on a grid. The resulting contour map was done on a clear Mylar sheet and resulted in a lot of squiggly contours over the face of the Utopia Planitia mosaic, but at least I had something semi-quantitative to show. There were areas of high abundance of hazards and areas of lower abundance of hazards. But the trick at that point was taking the landing ellipse on another sheet of clear Mylar, scaled according to the scale of the mosaic, and sliding that ellipse around until it fit, by eyeball estimate, in an area with the lowest number of hazard contours. Lo and behold, I found a spot where the ellipse looked to be in a fairly low hazard area, and it was not too far from a previously estimated place for one of the landing ellipses. So, it was more in the way of a confirmation of a previous candidate with some semi-quantitative tweaks to the positioning of the ellipse.

During the last meeting of the site selection and verification committee, Masursky and Carr took this map along with many other materials from the imaging team's efforts to the meeting. It was intense. After all, this was the do-or-die meeting. This was going to be the meeting in which a final decision was to be made. After a

RIGHT: Controlled photomosaic of Viking Orbiter images covering part of Utopia Planitia. This illustrates the type of data we had for landing site selection in the final days before landing. The ellipse is centered on the area's highest probability of landing for the selected target. The site was selected in part by the author after frantic landing hazard analysis while the orbiter and lander were still attached and in orbit, awaiting certification of a landing site.

Viking Lander 2

LEFT: First image taken by the Viking 2 Lander on the surface of Mars looking down at one of the landing footpads. This scene was up on the monitors at JPL the morning after the last landing site certification meeting and was the first evidence to the author that the landing had been successful. The rocks in the scene contain abundant gas bubbles, known as vesicles to geologists. This is a very familiar scene to anyone who has ever walked across an older lava flow on Earth and this looked shockingly identical to the rocks the author had walked on just a few weeks earlier in New Mexico.

decision was made, the commands would be sent up to the spacecraft that very night and the Viking 2 Lander would separate from the orbiter and descend to the surface. This gave the last meeting a certain immediacy and anxiety. During the meeting the imaging team results were good, and my map looked interesting to the tense room, but the level of uncertainty about the actual surface conditions still remained a bit high. I recall that most of the Big Guys felt that the Utopia Planitia site was covered in part with dunes and that the presence of dunes would mitigate some of the rough terrain. But I confess that I never saw anything that looked like dunes to me. Since I could not easily assess where there were dunes and where there were no dunes, I had simply ignored them in my analysis and focused on bigger things that I could confidently identify.

After many presentations and going back and forth on the arguments for different results, it looked like the rock-abundance arguments were going to win the day. At least during the final informal or straw man vote the thermal inertia results appeared to be the winner. There was one problem: the thermal inertia results from Hugh Kieffer looked good in a few places, but there were no images in those locations. During all this, Jim Martin did not commit to any site. After the closing informal vote, Jim said thank you to all who had presented opinions and asked that everyone but the voting members leave the room so the site certification and verification committee members could discuss and finalize the results. The door was closed, and the rest of us went back to other projects that were cooking. I eventually left for the day without knowing the outcome of the site selection meeting. Nor was there any communication outside the operations about the outcome of the landing that was to take place that night.

On entering the orbiter imaging team work area the next morning I thought that it would be obvious if the landing had been a success. But there were no leftover cigars or champagne glasses. Everyone that I could see was quietly working in their respective stations. Glancing around, I saw the monitor on the wall that showed the now familiar image of a landing pad and an adjacent field of rocks that looked much like the Viking 1 Lander initial surface images.

But on closer inspection there was something different. The rocks were different and in fact one big rock near the lander foot was pockmarked with gas bubbles that we geologists call vesicles and are typical of volcanic rocks. This was the first image from a very successful Viking 2 Lander! And there followed some initial segments of the local panorama.

It was a shocker. Rocks everywhere, including some big ones. We had successfully landed, but with some great luck. Some of those rocks were much taller than the clearance on the bottom of the lander. There were so many rocks that one of the footpads was perched on a rock, tilting the whole lander significantly. A bigger rock would have meant trouble up to and including a tipped over lander. We had made it to the surface but with a lot of luck in our favor.

It was only later that I learned that the landing site was pretty much right where I had placed the landing ellipse. I heard a rumor that I never confirmed, that at the end of the closed-door landing site selection meeting, Jim Martin pulled rank of a sort and liked where I had placed the ellipse. The rumor was that while the remote sensing results on rock abundance were mostly convincing, and in hindsight actually pretty accurate about the fact that there were too many rocks in adjacent areas of the proposed target site, he felt that ultimately he preferred the visual site locating method, and my map provided at least some basis for where to put the ellipse in the available target region. I'll probably never find out if this was true, and I may be biased to believe my map was significant in the process. Besides, Jim was a proponent of the dunes hypothesis that I had not included in my analysis, so whether my map was just a bit of additional evidence or not is up for debate. In any case, the site was where I had moved the ellipse. Big thumbs-up there. Eventually, near the end of my summer tenure at Viking operations I got a nice mounted color image from the first lander that was signed by Jim Martin with a little note thanking me for my participation, as did many of the other summer students who worked on the Viking Project.

LARRY — THIS HISTORIC FIRST COLOR PICTURE OF MARS WAS TAKEN BY VIKING LANDER I ON JULY 21, 1976, AND IS GIVEN YOU IN APPRECIATION OF YOUR WORK THIS SUMMER FOR THE VIKING PROJECT. BEST WISHES IN YOUR FUTURE CAREER.

James Martin
30 SEPT 76

ABOVE: Certificate of appreciation presented to the author at the end of the summer of 1976 and signed by Jim Martin, manager.

GEOLOGIC EXPLORATION OF THE NEW WORLD FROM ABOVE

WHILE THE LANDER TEAM SET ABOUT doing their thing and analyzing the scene and local soils around the second lander, we on the orbiter imaging team returned to the exciting task of exploring the rest of Mars from orbit. And there were some exciting new results coming in. We were getting high-resolution mosaics of many things that were originally seen in the Mariner 9 images from several years before, including large sinuous river valleys and scoured outflow channels, many areas of beautiful windblown features and dunes, and of course the volcanoes. But we were also beginning to see things that were not obvious in the lower-resolution Mariner 9 images.

The flood of new data from the orbiters kept the orbiter team busy. I recall at one point sitting at the big conference table in the Viking orbiter team work area, sorting through some recently arrived images while working on a map, when Jim Gooding, one of the graduate students working with Klaus Keil over in the Viking Lander team room, popped through the door and was moving around looking at all the images on the wall from the orbiters. As he was looking over my shoulder at all the new images I had laid out he suddenly exclaimed, "Wow! You guys have lots of new things to look at! Over on the lander we have to stare at the same thing every day. There is nothing new."

This is where I really felt gratitude in ending up on the orbiter team rather than the landers team. The landers were the media stars, but the real meat of exploration was taking place on a global scale with the orbiters. We of course felt somewhat sorry for the poor slobs stuck on the ground rooted to one place with very little in the way of new things to look at and analyze. It was all great fun the first day. But then they woke up the next day and there they were, looking at the same thing. Of course, they were doing a variety of interesting experiments measuring the composition of soils and the tests for traces of life. But the scenery never changed much.

Meanwhile the list of new and never-seen-before things in the orbiter images was long, and it was difficult to keep up as the data came in each day. Even things that we thought we had seen before on other planets were turning out to be new

> The landers were the media stars, but the real meat of exploration was taking place on a global scale with the orbiters. We of course felt somewhat sorry for the poor slobs stuck on the ground rooted to one place with very little in the way of new things to look at and analyze.

revelations of one type or another. One thing that was coming into focus was the fact that many craters on Mars were very different from the ones we were accustomed to seeing on the Moon. The material thrown out from craters during the impact process, known as ejecta, was even different. Surrounding many Martian craters we were seeing enormous splash-shaped patterns with sharply defined margins. The edges of the ejecta were escarpments or ramparts, completely unlike the ejecta of lunar craters. It appeared like the pattern one would expect for an impact into something more cohesive and wet, somewhat like the pattern when you throw a rock into some stiff mud.

This prompted a little project wherein I started assembling a catalog of many examples of this new crater type and ultimately led to my being included as a coauthor on a paper about rampart craters, as they were so-named by Mike Carr because the margins of the ejecta ended in steep slopes rather than the feather-edged extremities

BELOW: Mike Carr, principal investigator and head of the Viking Orbiter imaging team, in the orbiter team work area grabbing a few moments for some science analysis of an orbiter mosaic, 1976.

I like to think that it was the prevalence of lava flows on planetary research that initiated this interest in the physical characteristics of volcanism.

of ejecta that we had seen on all craters on the Moon. Rampart craters, or "splosh craters" as some referred to them, were an interesting subtopic in the greater field of impact crater studies for many years. Apart from the study of how the impact process worked, many studies to follow showed that there was a tendency for these rampart craters to occur in higher latitudes, suggesting to everyone that maybe the surface there was rich in ice or water, giving the ejecta a denser, splashy characteristic instead of a dry spray of broken material. This was one of the first of many indications that the surface of Mars, while dry and desert-appearing, may in fact be more like some arctic permafrost with lots of water ice buried below the surface.

The new information coming down from the Viking Orbiters regarding the volcanoes of Mars was also revealing a new world that was only partly explored with Mariner 9. These were heady days when each day might reveal some new detail about the great volcanoes of Mars or details previously unseen on the smaller volcanoes. It was a feast for the eyes for many of us interested in volcanism. Mars was proving to be a veritable museum of volcanic landforms. And then there were volcanic terrains, especially vast lava flows, and the summit calderas of volcanoes that promised to not only revamp our understanding of Mars volcanism but would expand our geologic knowledge about many characteristics of volcanism in general. I was so intrigued by the lovely shapes of long streaming lava flows that I recall seeing analogies to lava flows in many places.

One hot summer day during the mission, as I was walking across a parking lot at JPL, I noticed that the tar along cracks in the asphalt was streaming away from the cracks down the sloping parking lot surface, looking like lava flowing from a volcanic fissure. This got me thinking how it would be useful to explore what the different shapes of lava flows meant in terms of the viscosity and other factors related to how volcanic eruptions took place. I was never able to pursue that train of thought, but others later did, spawning a whole field of the study of the physics of viscous fluids flowing on surfaces. Later researchers, like volcanologist Steve Baloga, took up the challenge and wrote some of the first papers about the fluid flow and heat transfer in lava flows and some of the controls on their emplacement.

The field of lava flow research expanded in the years following and included the likes of famous volcanologist George Walker who proposed that the rate of eruption was as influential on lava flow length as was the viscosity of the magma. And there were many more studies that followed. Now, many of us engage in a variety of research sorting out the details of how lava flows work. I like to think that it was the prevalence of lava flows in planetary research that initiated this interest in the physical characteristics of volcanism. After all, with planetary studies we were forced to use the morphology of geologic characteristics to get at the meaning of the landforms being observed. In a sense, planetary geology had reinvigorated the old science of geomorphology, which was more concerned with the origin of landforms than the mineral and chemical composition of rocks that was much more the standard form of geological research.

The Viking Orbiters took images of the planet in long swaths of multiple overlapping images or frames. Because each orbiter camera system consisted of two cameras, an A camera and a B camera, pointing at adjacent areas on the surface, these swaths consisted of not one but two side-by-side strips. The result was a series of long mosaics in which the cameras stepped across the surface capturing high-resolution views of targets of interest. In addition to the early mosaics of the landing sites, as the mission progressed mosaics were captured of many terrains of interest previously identified in the Mariner 9 images, but now much more clearly resolved in the Viking data. Among these were features such as the enormous outflow channels, long valley systems that appeared to be ancient dry rivers, the vast Valles Marineris canyon, dune fields, polar terrain, and, of course, the volcanoes. I set about collecting as many of the volcanic images as were available in the early weeks of the mission.

About this time Jayne completed her field mapping back in New Mexico and was sent by Wolf as his second student to JPL to also work with the Viking Orbiter team as a graduate student intern.

She had her own projects to work on with the various Viking team members, including a small crater study with Jim Cutts. But as a personal project, together, we set about exploring this new world of volcanic features that Viking was revealing and quickly narrowed our attention to studying some of the new images of the large volcanoes. One series of image mosaic swaths that captured our imagination included the summits of three volcanoes in the vast Tharsis Montes volcanic area, Ascraeus Mons, Pavonis Mons, and Arsia Mons. These were the same volcanoes that were prominent in the first Mariner 9 images as the global dust storm gripping the planet in the early days of that mission. All three mosaics crossed the volcanoes in a line from southeast to northwest, showing details of the flanks of the volcanoes and summits, so they represented nice "geotraverses" across the main volcanoes and their surroundings.

ABOVE: Scene from the Viking Orbiter imaging team work area in Building 264, JPL. The author's wife, Jayne, doing crater counts on an early Viking Orbiter image mosaic. Analysis was all about light tables, physical image products, and transparent overlays at that time.

Each volcano was similar in that they all had enormous summit calderas surrounded by concentric faults that represented the effects of stresses created in the surroundings during the collapse of the calderas. Because there is no erosion, at least nothing like the wet and rainy surface of Earth, these sorts of details are perfectly preserved on Mars. This allows us a peek at the processes associated with fundamental volcanic phenomena not generally available on Earth. Not only were caldera details on display, but a vast array of lava flow details and other associated features were available for analysis. This preservation of fundamental volcanic features was an exciting new area of study for many of us fascinated with volcanology. After the end of our stay at JPL and on returning to the university at the end of the summer of 1976, we had amassed enough information on these three Tharsis Montes volcanoes that we decided to publish a paper based on our results. The final work appeared in the planetary journal of choice at the time, *Icarus*, and it was the very first of our many joint papers on planetary geology topics that followed over the years on our research from Mars to Venus.

Before the Viking 1 Orbiter mission ended on August 17, 1980, and Viking 2 Orbiter mission ended on July 25, 1978, 97 percent of the surface of Mars had been imaged at better than 100-meter resolution, and in some images at better than several tens of meters. For the first time since humans found themselves wondering about the mysterious red planet, we now had the entire surface of Mars laid out before us and most of its geologic history laid out for inspection and interpretation. The Mars of the imagination began to evolve to a Mars of the scientific investigation buoyed by the science triumvirate of observation, theory, and testing. Entire books have been filled with the scientific findings of the Viking Project and it would be an ambitious and perhaps exhausting textbook enterprise to do a thorough examination here. But we can review a few of the basic things that we had learned about the new world up to this point and introduce a few of the new questions that we got in return from our eyes in the sky looking down at Mars.

> For the first time since humans found themselves wondering about the mysterious red planet, we now had the entire surface of Mars laid out before us and most of its geologic history laid out for inspection and interpretation.

FOUR

New World, New Globe

"For there is a new place for those who are willing, who are able, who are strong. We are going west. There is, I believe, a new world somewhere waiting. The moon that shines here will shine there, but here the land is broken and there it is whole."

—KATHRYN LASKY, *Spirit Wolf*

TO BE A TRUE MARTIAN, IT is necessary to be familiar with Mars geography, that is, the locations of features on Mars and their names. There are many Mars scientists who have worked on a particular problem related to the overall scheme of Mars exploration yet know very little about Mars as a whole, and certainly cannot tell you where all the wonderful places are located. Once you familiarize yourself with the land, you can start your own itinerary. But good luck, the total area of the Martian surface is similar to the total land area of Earth. There is a lot to see.

RED PLANET RENAISSANCE: THE BIRTH OF MODERN MARS GEOLOGY

BY THE END OF THE SERIES of Mariner and Viking missions in the late 1970s, the old mysteries of the telescopic era of Mars exploration were finally being scrutinized and revised. We had developed a good understanding of the physical characteristics of the Martian surface that told us in many ways that Mars was actually an entirely different planet than we had expected. We had a new world and a new global geography, or aerography as it is sometimes called. The "new" Mars was not less exciting but actually more exciting, just in a different way than initially speculated in the days before spacecraft were sent there.

We learned very early, during the Mariner 9 mission, that the classic albedo features of telescopic observation were largely unrelated to the physical topography of the surface. Then we learned that there was an odd global dichotomy, or two faces to Mars. From Mariner 9 results we had learned that, at the global scale, Mars's topography was divided into two distinct hemispheres, a northern hemisphere that consisted of low plains and a southern hemisphere that was elevated and more cratered. The dividing line between the two different hemispheres is not quite at the equator, but is instead slightly tilted so that it crosses into the northern latitudes part of the way around Mars and dips closer to the equator on the other side. The southern half of the planet's surface is above the average surface elevation and the northern half is below

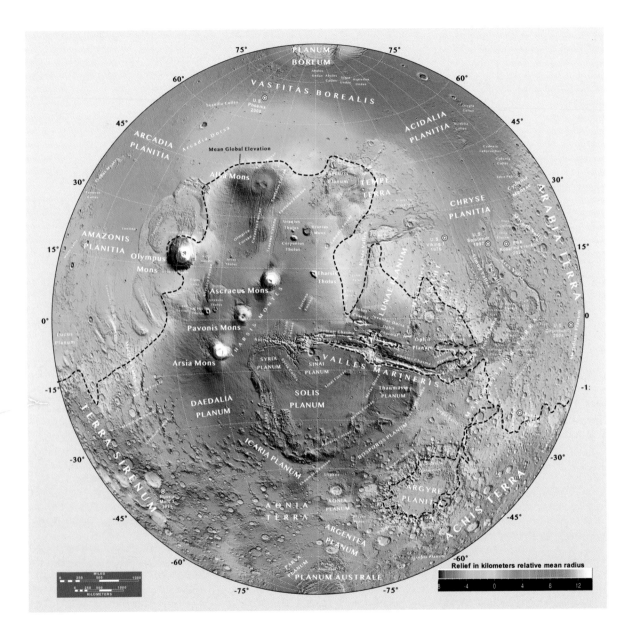

ABOVE: Western hemisphere of Mars.

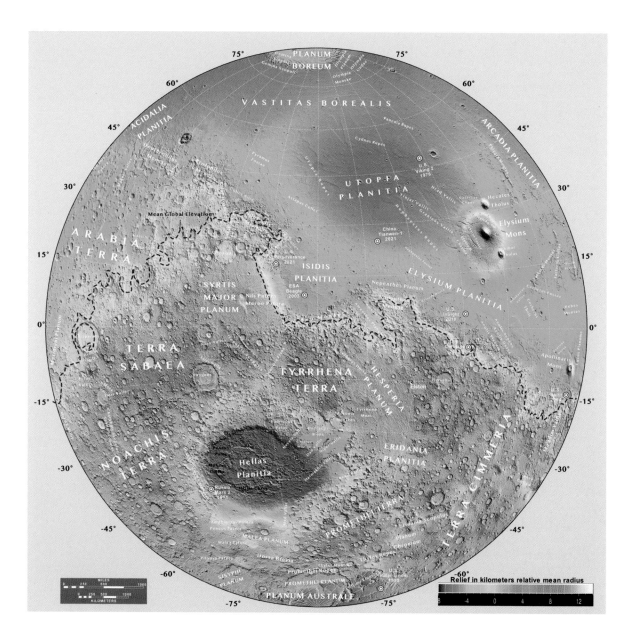

ABOVE: Eastern hemisphere of Mars.

the average surface elevation. This average elevation is used as the Martian equivalent of "sea level" and all elevations on Mars are measured up or down from this elevation level. And the "sea level" on Mars pretty much follows the boundary between the two hemispheres, as you might expect. Abundant craters characterize the terrain of the higher southern hemisphere. This was the part of Mars that was seen in the first three Mariner missions and had caused the impression initially that Mars was just another cratered planet. On the other hand, plains generally consisting of broad areas with fewer craters than the highlands dominate the terrain of the lower northern hemisphere. But in detail both hemispheres contained geologic complexity that went far beyond that simple global view.

We had a new world and a new global geography, or aerography as it is sometimes called.

The origin of this dichotomy is unclear, although there are many candidate explanations, called hypotheses in science, that fit what we know and thus provide some directions for future observation and testing. It could be that Mars, like the Moon's near and far side, had several large-impact basins on one hemisphere (in this case the northern hemisphere) and that those overlap and have subsequently been eroded and partially filled with volcanism. Or maybe internal forces were asymmetric and rebuilt the crust thinner in the northern hemisphere. There are other ideas, but one thing is clear: the dichotomy is part of Mars's deep past and getting at the answer will require getting down on the ground and pounding on the rocks and doing so in enough places to build up a list of evidence that can be used to test these ideas.

But within those two different hemispheres we had also inventoried many new and exciting geologic landforms that represented records of Mars's very active geologic history. The menagerie of geologic landforms was extensive. There were volcanoes and lava flows, what appeared to be water-cut channels, and areas of vast outwashes, canyons, all sorts of fractures and fault lines, rows of pits in a line, clusters of small hills, long ridges, long cliffs, windblown and sculptured features, streaks and dunes, large areas of accumulated sedimentary layers, and of course polar ice caps and related features. The list is even longer, but this gives the notion that there is a great diversity of landforms as befits what is obviously a geologically complex planet.

A TOUR OF THE MARTIAN LANDSCAPE FROM ABOVE

TO PROPERLY READ A MARS MAP, you need a program to recognize the players. The distinctions between some of these features can be academic, but many of them are of general interest and common enough that, as a start and a guide for exploring Mars maps, we can briefly outline a few of the alien-sounding names of Martian terrain. There are many other feature types than those listed here, but many of the most common refer to large features more likely to be encountered when looking over a Mars map, or are important in the narrative that follows, including:

> "Geography is an earthly subject, but a heavenly science."
>
> —EDMUND BURKE

- **Catena** (plural: *catenae*). A linear string of craters, either impact craters or pits. Pits sometimes occur along fractures or may be arrayed along the floor of a valley bounded on two sides by straight fractures.

- **Cavus** (plural: *cavi*). A depression with irregular margins. The irregular shape and absence of raised rims are not typical of impact craters, so these appear to be something else.

- **Chaos** (plural: *chaoses*). An area where the surface is a jumble of hills and scarps, sometimes in a gridded pattern, sometimes not.

- **Chasma** (plural: *chasmata*). An elongated depression/trough with steep sides.

- **Collis** (plural: *colles*). A small hill or knob. Usually these are clustered and, hence, colles on a map.

- **Crater.** A circular depression, often with raised rims created by an impacting body. The proper names assigned to larger Mars craters are derived from the names of prominent scientists who had contributed to the study of Mars. Craters smaller than sixty kilometers are named after villages of the world with populations of less than 100,000.

- **Dorsum** (plural: *dorsa*). A ridge or an elongated prominence.

- **Fossa** (plural: *fossae*). Ditch. Long, narrow, shallow depression. They generally occur in groups and are straight or curved.

- **Labyrinthus** (plural: *labyrinthi*). Valley complex. Intricate, intersecting valleys, of which there is only one example, Noctis Labyrinthus.

- **Mensa** (plural: *mensae*). Flat-topped prominence with cliff-like edges.

- **Mons** (plural: *montes*). Mountain. A large topographic prominence or chain of elevations.

- **Patera** (plural: *paterae*). An irregularly shaped crater, with low relief that has scalloped edges, without raised rims in many cases, and sometimes with radiating channel-like features. It is not interpreted to be an impact crater.

- **Planitia** (plural: *planitiae*). Plain. Smooth, low area.

- **Planum** (plural: *plana*). Plateau. Smooth, elevated area.

- **Rupes** (plural: *rupēs*). A cliff or scarp that is straight or linear rather than sinuous.

- **Terra** (plural: *terrae*). An extended areal region or landmass. It is used in reference to the older, cratered highlands.

- **Tholus** (*plural*: *tholi*). An isolated steep or dome-shaped small mountain or hill.

- **Unda** (plural: *undae*). An area of dunes that are very wavelike in appearance.

- **Vallis** (plural: *valles*). A sinuous valley. These have the appearance of water-carved valleys. These are generally named after the word for Mars in various languages.

- **Vastitas** (plural: *vastites*). Extensive plain.

Then there are a host of smaller features or features that are less common:

- **Fluctus** (plural: *fluctūs*). A flowlike feature. Only one example of this type of feature is found on Mars, Galaxias Fluctus.

- **Labes** (plural: *labēs*). A feature that has the appearance of a landslide. The Latin meaning for labes is a "falling in" or "sinking in."

- **Lingula** (plural: *lingulae*). Tongue-shaped feature.

- **Palus** (plural: *paludes*). A dark plain.

- **Scopulus** (plural: *scopuli*). A cliff or scarp that is irregular or lobate in appearance.

- **Serpens** (plural: *serpents*). A sinuous ridge.

- **Sulcus** (plural: *sulci*). A feature that has the appearance of a furrow, ditch, or wrinkle. These often occur in groups.

Those are long lists and they are also somewhat confusing and academic lists of strange terms. But they do represent an attempt to assign feature types to the post–Mariner 9 and Viking landscape menagerie. Although the names appear scientifically obscuring, they include some features that are worth taking a look at from the context of what we have determined about them in a geological sense. What was the significance of those features for understanding Mars? And what did we really know about them at this post-Viking mission point in the long-term exploration of Mars?

THE LANDSCAPE OF MARS: GEOLOGY ILLUSTRATED

STARTING WITH THE MORE COMMON OR larger feature types, several are at the top of any Mars tourism list, or at least, if you look at a modern map of Mars or any random image of the surface of Mars, you might encounter them. For many casual Mars tourists, the features to see might include features that are interpreted to be interesting geologic features, such as impact craters, volcanoes, water-cut channels and floods, fields of dunes, and all sorts of faults and fracture features. There are many other surface features that are not so obvious from their shapes alone and perhaps more picturesque, and we will take a look at those later as the exploration unfolded. But here is a sampler of some of the standouts that we had on the plate of exploration in the post–Viking Mars era of exploration. To help further organize the list in one's mind, we can collect many of them into their likely geologic families. So

> "I am here tracing the History of the Earth itself, from its own Monuments."
>
> —JEAN ANDRÉ DE LUC

now we are going to talk about some of these features by giving them the basic geologic interpretation. For example, we generally interpret a mons as a volcano on Mars, not a giant ant hill. There is of course a caveat to doing this, namely, that it is a somewhat dangerous exercise given that the names were originally developed in order to avoid blanket geologic interpretations. Sometimes the geologic interpretation I mention will not work, but the great majority can be so lumped, so we will proceed with that caution in mind in an effort to bring some control and meaning to this otherwise daunting array of geographic terms.

We can start by stepping outside the alphabetical order of the previous list and organizing these in order of their prominence, either their size or the frequency with which they are encountered when looking at the global maps of Mars. And then we can start talking about collections of features that are generally, but not exactly, formed through a similar geologic process.

Planitia. The largest feature types are regions of similar terrain. Planitia are low and plain-like areas. As an aside, the more common pronunciation is the classical Latin (plaˈni.ti.a) as opposed to the ecclesiastical (plaˈni.t͡si.a). I have heard both pronunciations used, but the former appears more common among planetary geologists. This only serves to illustrate that almost everything becomes complicated when you attempt to be a bit more rigorous. Planitiae (remember the plural) appear featureless in large maps and on global views of Mars. But up close they too contain many craters, and the craters are smaller, which means that the surface has not collected as many craters over time, and the craters appear more rugged and are thus generally much younger than those in areas like the highlands. The low northern plains of Mars are primarily planitia, and planitiae are also present in the floors of some of the larger old impact basins such as Hellas Planitia. But as we have seen in the discussion of the Martian global dichotomy of elevations, the northern plains to a large extent may also be the remains of overlapping large-impact basins as well.

Chryse (CRY-see) Planitia is an example on the edge of the northern lowlands abutting the highlands to its south. It's important in the history of Mars exploration because it was the site of the first successful lander, Viking 1, and, about twenty years later, lander/rover, Pathfinder/Sojourner. The thing that made Chryse Planitia attractive for a lander was, first, the relative flatness and, second, the fact that it was low. The flatness permitted reduced risk of landing on something rugged. Low altitudes were better because the elevation provides a bit more atmosphere for slowing down before arrival at the surface. But there was one other thing about Chryse

Planitia of importance to the Viking mission goals: the fact that a whole series of large flood or outflow channels appeared to empty into this low-lying area. In those days one of the concepts that drove landing site selection was the desire to have access to as many rock types as possible and it was thought that an area that had been inundated with the debris washed down from adjoining terrains provided the potential for a variety of rocks. Also, Viking goals included a search for life. A place where water had once flowed across the surface was a place where life may have had some chance of occurring. And if it was not present there at some point in time, then materials potentially in contact with life may have been swept in from upstream, increasing the chances of finding something. In the early days this was called the grab bag concept, meaning that instead of landing on just one rock type, the landing site might be a deposit at the end of a large outflow that swept up all sorts of different things along its path and piled them up at the terminus just waiting for a stationary lander to come along and sample.

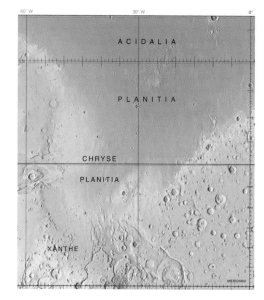

ABOVE:
Color-coded relief map of the Chryse Planitia and Acidalia Planitia areas of Mars, site of the Viking 1 and Pathfinder landers.

One would think that, being low and flat, the planitia would be boring, but there was plenty of evidence to suggest otherwise. For example, the surface around Viking 1 Lander was littered with rocks, some large, and a rolling terrain out to the horizon. So the planitia were clearly not competition for the Bonneville Salt Flats in Utah. The other thing was the presence of a variety of enigmatic landforms. Acidalia Planitia adjoining Chryse Planitia to the north is a good example. The surface of Acidalia Planitia contains many features that are interpreted as evidence for the influence of subsurface ice, a kind of Martian permafrost. In a few places there are collections of domical hills, many with small "craters" on their summits.

Other features on the lowland planitia of the north included various types of "patterned ground" common in some arctic regions of Earth. Then there are later studies of the mineral composition of the planitia of the north that provide evidence for minerals that are formed when rocks are altered by water, including minerals such as clays. The northern lowlands as a whole are also known to have volcanic rocks of somewhat different chemical composition (what geologists call andesitic rather than basaltic). Whether this was because the original lava flows that covered those plains

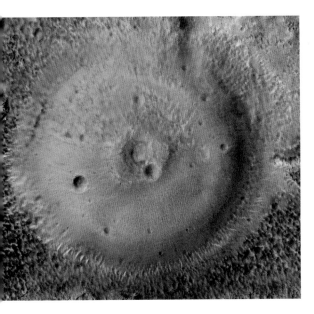

ABOVE: Unusual pitted mounds in the high latitudes of Mars such as this one over 100 meters across in Acidalia Planitia are interpreted as possible "mud volcanoes" driven by subsurface ice and gases. One of the gases could be methane, which is discussed in a later chapter.

were more andesitic or a result of later alteration (or chemical change) due to the presence of water is still being debated. Finally, there have been many research studies about the possibility that the northern lowlands were once a vast Martian sea, including efforts to trace shoreline "bathtub rings" and even tsunami effects on the margins of the plains created when large-impact craters impacted in this hypothesized ocean. These are ongoing areas of investigation and speculation. But the idea that water oceans at one time inundated the northern lowland planitiae has been gaining traction in Mars scientific circles.

Hellas Planitia in the southern hemisphere is the lowest elevation on Mars and is the floor of the largest unambiguous impact basin, sometimes referred to as the Hellas impact basin. Its relatively circular shape and bright, dusty surface was such that it was frequently seen in early telescopic observations. Recall that in the earlier telescopic maps it was assumed that bright areas were like continents and dark areas were low seas. So Hellas was initially thought at that time to be a high elevation. Coyote Mars successfully did the old switcheroo on us again since Hellas Planitia is in fact now known to be the lowest place on Mars.

Hellas Planitia is a kind of Death Valley on Mars. The elevation of the floor is nearly seven kilometers below the global mean elevation, so the atmospheric pressure here is higher than anywhere else on Mars. The pressure is sufficient that ice and water could be stable at the surface under certain conditions. The terrain on the floor of Hellas Planitia is a complex of unusual landforms, many of which are interpreted to be results of the flow of ice, both laterally and in some places possibly upward as domes of ice rising from depth.

There is another thing about the low elevation here and the correspondingly greater atmospheric thickness, and it was highlighted with the results of an instrument on the Mars Odyssey orbiter, the Gamma Ray Spectrometer. Without getting too deeply into that instrument and that orbiter mission at this point, the purpose of the Gamma Ray Spectrometer was to map the elemental composition of Mars's surface by using the energy backscattered from incident cosmic rays. Diverse elements respond differently and the difference can be detected from orbit. This is a clever use of the fact that the surface is exposed to all sorts of incident cosmic rays, unlike Earth, where both the thick atmosphere and the magnetic field largely shield the surface from much of the cosmic ray bombardment that it would otherwise receive. Because Mars does not have a protective magnetic field to deflect solar particles and other cosmic rays, the Gamma Ray Spectrometer worked pretty well. In addition to identifying many elements, which could be used to estimate various types of minerals and volatiles, such as water ice, it also showed that, although the radiation at the surface of Mars is significant everywhere, it appears lowest at the floor of Hellas Planitia. One conclusion from this is the fact that future human explorers would find themselves a bit more protected from cosmic radiation on the floor of Hellas Planitia. In general the radiation environment on the surface of Mars is an area of considerable concern for future exploration.

Planum. These are a sort of inverse of planitiae, that is, instead of being low plains they are broad elevated plains. Several examples occur around the Tharsis region that surrounds the numerous volcanoes there. Other plana include volcanic areas covering parts of the older highland terrains such as Syrtis Major Planum and Hesperia Planum. These are lava plains for the most part, although an exception includes Meridiani Planum, a flat-lying relatively smooth area in the lower reaches of the cratered highlands. This was the site of the Mars Exploration Rover Opportunity traverses where the

BELOW LEFT: The floor of Hellas Planitia, one of the larger impact basins and the lowest surface on Mars. Because of the high southerly latitude it is thought that the complex terrain here is in part a result of ice and erosion processes.

BELOW RIGHT: Color-coded relief map of the elevated plains of Solis Planum and Daedalia Planum.

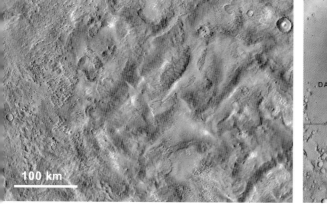

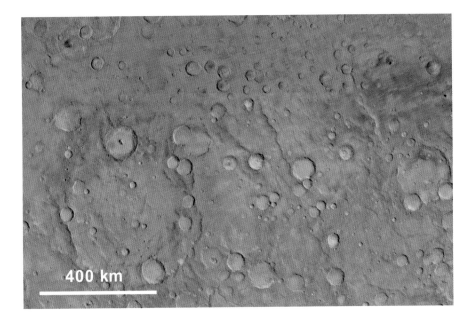

400 km

ABOVE: Viking Orbiter–based mosaic of the dusty and cratered highlands of Arabia Terra.

surface is covered with fine sands rich in the mineral hematite and cemented with sulfates. This simply serves to reiterate what was said earlier, namely that the fact that the geographic terms for different terrains are not a substitute for a geologic origin. In this case, being a flat surface that is also an upland does not always imply that it is a lava-covered upland. We will return to Meridiani Planum later for an up close and very personal view of this area. Another planum that is not an elevated lava plateau is Planum Australe, the south polar cap, which is also surrounded by a series of other plana that are essentially old cratered highlands seriously eroded or covered by polar-related deposits. The north polar cap is Planum Boreum.

Terra. Large expanses of what are now interpreted as the older cratered highlands are generally named as terra. Noachis Terra is a prime example. Located in the heart of the cratered highlands west of Hellas Planitia, Noachis Terra is populated with many large-impact craters and the craters are generally very flat, giving the appearance of considerable erosion, and hence rather old. In fact terra such as Noachis are among the oldest areas of Mars. As such they contain a record of some of the earliest preserved events on Mars. This association played a role in the naming of Mars geologic ages, which we will visit later.

Arabia Terra is another broad region of older cratered terrain, this one being west of Isidis Planitia and the part of the ancient cratered terrain that extends into the northern hemisphere. It is the most densely cratered part of Mars. This region appears in large maps as a broad, boring area of old cratered terrain, but as with almost everywhere on Mars, closer inspection reveals all manner of interesting things.

Craters. With a few exceptions, one of the most common features on all planets outside of Earth are craters created by large objects impacting planetary surfaces. Craters are usually round or roundish, and they usually have raised rims. As a result, craters are fairly obvious, and it is agreed there is little risk of misinterpretation regarding their origin. That means that craters are one of the few features that have geographic names that agree with the geologic name for the feature and there was no real need to give them a Latin term like all the other landscape features.

The study of impact craters is an entire field of study in itself and includes studies of examples on Earth, the planets, and many of the moons of the planets. Craters on Mars have presented some new wrinkles. We had seen craters before on the Moon obviously, but Mars craters were noticed in the first images from Mariner 4 as somewhat different in appearance from lunar craters. For example, the first images from Mariner 4, the images that revealed the then depressingly cratered surface of Mars, were also somewhat enigmatic because the craters were low in topographic relief and the floors were flat and featureless, a far cry from the rugged, steep-walled, and central peaks of even large craters on the Moon. Additional images with Viking and later missions also provided evidence that the rims of many of these otherwise flat craters on Mars had peculiar scallops and valleys. We later realized that they appear somewhat subdued compared with places like the Moon probably because, unlike the Moon, there has been an active atmosphere and a long history of erosion on Mars over billions of years. And there are other considerations, including the presence of ground ice and rock alteration due to water that has no doubt enhanced the ability of erosion to wear many craters down. The Martian highlands are now recognized as being an older terrain where the erosion has had plenty of time to act.

There are also craters that are not so old and preserve some other unusual characteristics. Like the Moon, we see on Mars that small craters, less than a few kilometers in diameter, tend to be more bowl-shaped. As the size transitions to larger craters, there is a more complex form with terraced walls and central peaks. Larger still, the central peak becomes a peak ring. At sizes approaching several hundred to thousands of kilometers, dimensions that are referred to as an impact basin, there may be concentric rings. The deepest areas on Mars, including Hellas Planitia, are all formed from some of the largest impacts, probably very early in the history of Mars.

But beyond this differing morphology depending on their size, which we had also seen on the Moon and other planets like Mercury, Mars also has an array of impact crater morphologies that is more diverse. A relatively new class of crater was identified on Mars during the early days of the Viking Orbiter mission when a peculiar petal-like pattern of ejecta was noted around a certain class of Martian craters. We had not seen these on the Moon, so it appeared that there was something particular about Mars that led to their formation. So we cobbled together the first paper outlining their identification, characteristics, and possible origins. I must have seen too many cladograms showing the morphologic groupings of fossils somewhere in my distant past geologic education, because I had somehow got it in my mind to do a series of images of these craters followed by simplified sketches showing the different types of these petal ejecta or rampart craters and how they formed a continuous range. The chart started with normal appearing craters and stepped through craters with increasingly more ornate ejecta patterns. While my artwork did not make it into the paper, I was very proud that, as a Viking intern, my name made it into the list of Viking team authors right along with the likes of the first author, Mike Carr.

As we will see later, craters serve another purpose beyond studies of the impact process. Early in planetary exploration during the days of initial lunar exploration, it was obvious that some areas had more impact craters than other areas. Since craters do not all form at the same time but occur over long periods of time, it was also obvious that the more craters a surface contained the longer the surface had been sitting there, accumulating impacts. Ergo, the older the surface, the more impact craters that had accumulated. If anything happened to the surface, such as erosion or volcanic eruptions, then the previous record of accumulated craters would be erased or partially erased, thus resetting the accumulation clock. This is the basic reason why the Moon has more impact craters than Earth. The Moon has had very little erosion or geologic forces in operation over billions of years, whereas Earth's surface is constantly resurfaced by erosion and geologic processes like volcanism and plate tectonics. Crater abundance therefore became the standard for determining relative ages and has served us well in putting semi-quantitative or relative "ages" on Martian surfaces and geological features even if we don't yet know the absolute age or date of formation.

Craters on Mars have been given proper names that appear on any map of Mars. Names for those larger than about sixty kilometers in diameter are selected from the names of famous scientists, especially those who had contributed to the study of Mars, and other people who were prominent in one way or another in the history of Mars. Names for craters smaller than sixty kilometers in diameter are selected from

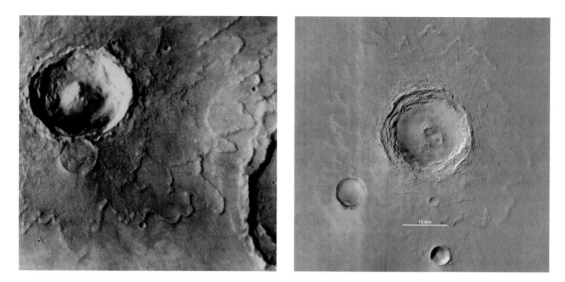

ABOVE LEFT: Historic original Viking Orbiter image of Yuty Crater located 2.4ºN, 34.2ºW. Yuty is a nineteen-kilometer-diameter crater, one of the first craters identified with unusual splash-like patterns of ejecta characterized by "ramparts" around the margins. **ABOVE RIGHT:** Santa Fe impact crater on Mars. Mosaic of MRO CTX and HiRISE images.

the names of small towns with populations of less than 100,000. The names used are somewhat more diverse in their significance, and the naming process can have a colorful history in some cases. For example, early during the Viking mission, some of the craters in Chryse Planitia, where the Viking 1 Lander arrived in the US bicentennial year of 1976, were given names in commemoration of important towns in early US history. Craters were given names such as Yorktown and Concord. But some sage realized that Santa Fe was an important town in early US history just like the early New England places, and Santa Fe was the name given to a beautiful, relatively uneroded, and otherwise classic twenty-kilometer-diameter impact crater a little to the south of the Viking 1 Lander site. Later on, I was fortunate enough to be co-author on the first geologic map, published by the U.S.G.S. Planetary Mapping Program, of the landing site area, including Santa Fe Crater.

Mons and Collis. Mons are generally volcanic in origin, but not always. For example, the Phlegra Montes is a range of rugged older terrain in the eastern margin of Utopia Planitia. But frequently the feature type appears in connection with big volcanoes. All the tall and really large volcanoes on Mars identified in the early days of Mariner 9 results are called mons. Four of the largest, Olympus Mons, Arsia Mons,

Ascraeus Mons, and Pavonis Mons, occupy an elevated region on Mars named Tharsis after the name for that area of Mars on the classical maps. Another large volcano, Elysium Mons, occurs in the Elysium region some four thousand kilometers to the west of Olympus Mons.

Olympus Mons is one of the more talked-about and spectacular features of Mars because, at between 600 and 700 kilometers wide, as large as many states in the western United States and countries in Europe, and 22.3 kilometers high, it is the largest volcano identified in the solar system. It is big and awesome, but there are a number of characteristics of Olympus Mons that go beyond these basic dimensions and make it even more impressive.

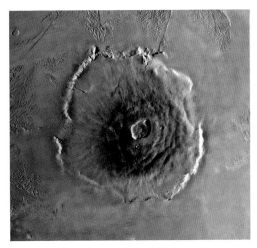

First, Olympus Mons is a type of volcano known as a shield volcano, which simply refers to the fact that the shape is like a warrior's shield lying on the ground, so it is not the type of volcano like Mount Fuji that comes to most minds when they think of volcanoes. But most people do know about the Hawaiian Islands being large volcanoes, and they are exactly the same type, shield volcanoes. Shield volcanoes are the result of lots of small lava flows accumulating from a source near the center. Steep-sided volcanoes that are the image we all have when we think of volcanoes are the result of somewhat more viscous lava flows and eruptions that produce lots of ash and debris, both of which accumulate in large volumes near the vent, so the whole volcano is a pile of many different types of volcanic materials, or a composite of lava and ash. The volume of shield volcanoes on the other hand is mostly lava flows, and they are somewhat less viscous or runny lava flows at that. So the overall accumulation is more spread out, flatter in the final landform, and without the steep-sided summits.

Second, on the top of Olympus Mons there is a series of overlapping volcanic craters, which geologists call calderas. Calderas like those on Olympus Mons are the result of collapse rather than enormous explosions blowing a giant crater into the volcano summit. While some calderas on Earth volcanoes form when an enormous eruptive explosion blows material out and forms a crater, collapses of the volcano

RIGHT: Famous Viking Orbiter image mosaic of Olympus Mons, twenty-two kilometers high and up to seven hundred kilometers across, it is the largest volcano in the solar system. It is an enormous shield volcano and one of the most famous landscape features of the red planet. Located at 18°N, 226°E in the Tharsis region.

OPPOSITE: Close-up of the Olympus Mons summit caldera.

summits are more often than not a result of changes in the volume of all that magma inside the upper parts of volcanoes. In big lava volcanoes like Olympus Mons the volume of magma decreases when it is extracted by an eruption. In some big calderas on Earth a similar thing happens when magma escapes as explosive clouds of ash and it is the exit of all that magma rather than the blowing off of the surface rocks that causes a giant caldera. With shield volcanoes, the magma comes out as lava flows mostly, so the collapse is somewhat more orderly. Rather than just sagging in as the magma is removed, because surface rocks are brittle, the sag usually takes place along encircling faults and the caldera floor is let down like a giant piston. But the process can repeat through the lifetime of a volcano and, because erupted volumes differ, the collapse can occur at different scales and centered on different places, creating a whole series of overlapping calderas. This is what gives the summit of Olympus Mons the appearance of a cluster of overlapping calderas. There were obviously multiple big eruptions during its later life and each one left its mark. And it is a big cluster of calderas; the width is sixty by ninety kilometers and three kilometers deep on average.

This means that the caldera width alone is larger than the width of some of the biggest volcanoes on Earth. Most of the Big Island of Hawaii would fit inside the caldera on top of Olympus Mons. If you stood on one rim, the far rim would be over the horizon. If there was an interstate highway system across the caldera, it would take you the better part of an hour to drive across it.

The feature type collis refers to any knob whether volcanic or not, but many of these are interpreted to be small volcanic centers. And because they are generally recognized only as clusters, for many years now the term has become used only in the plural, which is colles. There are several geologic features within this category, but perhaps one of the more intriguing is small cones clustered in areas of some young lava flows. Analogous features on Earth form when lava flows move across water- or ice-saturated ground. So it is thought that these may in many cases record a similar process that occurred during the eruption of these lava flows on Mars.

Chasma and Fossa. These features typically have the appearance of faults and structures that we know on Earth are created when the brittle rocks of the crust are broken by tectonic forces and earthquakes. Chasma are large, deep, often elongated valleys with steep, sharply defined margins, and fossae are particularly narrow and long crack-like examples of a similar geologic origin.

The Tharsis region is one of the big centers of fossa (faults and fractures) on Mars. If you look at a map of the entire Tharsis region there is a remarkable pattern

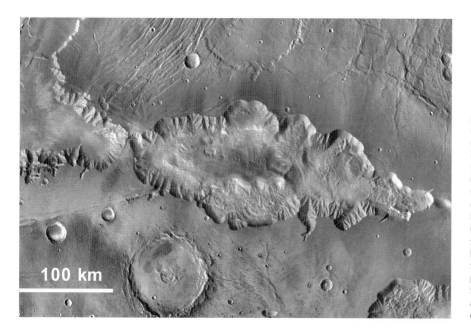

LEFT: Hebes Chasma is a closed depression five to six kilometers deep, 320 kilometers long, and 130 kilometers wide just northwest of the great Valles Marineris canyon. On the left is Echus Chasma, which forms the upper reaches of the enormous outflow channel Kasei Vallis. On the floor of Hebes Chasma is a large mesa-like surface, Hebes Mensa, over five kilometers high that appears to consist of layered materials, probably sediments and debris that covered the chasma floor and nearly filled the chasma at one time.

100 km

of straight fossa often extending for thousands of kilometers radially from the center of the region. Besides the largest four volcanoes within the region there are many smaller volcanoes and vast areas of associated lava flows concentrated near the center of the Tharsis area. We do not have all the answers, but one way this association of concentrated volcanism and faults and fractures may have come about is related to things that happen to a planet's crust when big things get built on top of it. All this volcanism confined to one region is a likely result of an anomalous amount of mantle melting, and some sort of broad mantle upwelling best accomplishes that at this scale at least. When mantle material ascends from the depths of the lower mantle there is not only excess heat, but the lower pressure causes the mantle to melt. This is somewhat like taking the cap off of an automobile radiator when the engine is warmed up. The sudden release in pressure moves the coolant in this case from a pressure where it is stable as a fluid to a pressure where that same fluid will be steam. In this case the reduced pressure causes it to become molten mantle. And thus, there is plenty of mantle melt, or magma, to supply volcanic eruptions, which brings us back to the main point. Because there have been so many eruptions supplying lava flows in the region, the lava flows piled up over Martian geologic history into a large bulge. Plus, hot mantle tends to be somewhat expanded, so some of the bulge probably comes from warmer mantle throughout the region. But all those lava flows and large volcanoes

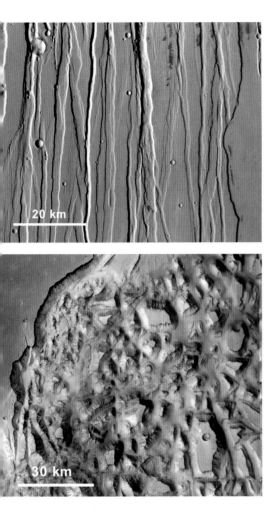

that were piled up on the surface have loaded the crust to the extent that it tended to sag with the weight. The result is all manner of radial fractures (fossa) and even some forces pushing outward and compressing the upper crust. And thus, the Tharsis region has created its own array of fractures, faults, or tectonic features.

Chaos. This is a distinctively Martian terrain type occurring over areas several hundred kilometers wide, many near the source area of large outflow channels and along the boundary between the highlands and lowlands or in the eastern reaches of Valles Marineris. It consists of an array of large polygonal blocks, often quite flat-topped with intervening deep troughs.

The overall effect is the appearance of a thick layer that has been broken and widened out between blocks. Although the pattern is somewhat polygonal, there is no systematic orientation over large areas and it appears chaotic, hence the name. This was an enigma because it is so unlike any terrain on Earth, but the fact that large outflow channels appear to be sourced full-blown in the vicinity of chaoses gave rise to the notion that perhaps melting of a deep ice layer, perhaps a frozen aquifer, at some point released vast amounts of water, both undermining the terrain above and sending torrents of water down the regional slopes into the lowlands.

Catena. This is a chain of craters that occurs on many planets and is the result of different processes. Some catena are simply an alignment of small-impact craters like bullet holes from a machine gun from material thrown out radially from a larger impact or a series of impacting objects. While there are examples of that process, most chains on Mars are rimless and roughly circular pits in the floors of fractures or fossae. This means that they are a process internal to Mars. They are somewhat exciting from a geological perspective because fossae are often long valleys with relatively straight margins in which the crust is dropped down between two or more parallel faulted margins. Such structures are call grabens by geologists. It is a common geologic structure and it is fairly common on Earth in areas where the surface is being pulled apart as a result of tectonic forces and the related earthquakes. But on Earth grabens develop over long periods of time and during that time erosion usually occurs, filling or partially filling the floors of the grabens while the margins that would otherwise be scarps are eroded into broader slopes. Sometimes the margins are very irregular and it takes a geologist to recognize them. Also because the floors are usually covered with the debris associated with erosion, any details of the chunk of the crust that has been dropped down are buried and out of sight. So on Earth there has not been much thought given to the geologic structure that may be present on the floors of grabens. On Mars most grabens are young enough that the limited erosion in the late Martian geologic history has allowed stark preservation and those details are perfectly preserved right out in the open for anyone to see. The presence of chains of pits on the floors of these Martian grabens was intriguing. Tractus Catena just north of Ascraeus Mons and southeast of Alba Mons is a good example.

This is a region with a series of north-northeast-trending grabens, many of which are part of the vast fracture system that is roughly radial to the Tharsis region. Little is known about the exact mechanism of formation, but it is postulated that as cracks develop along a zone that is extended, the resulting underground cavities allow the surface to collapse or drain inward into the void left behind during extension. That process is never allowed to run to full completion on Earth where water erosion acts continuously to fill and shape any low depression. But on Mars the arid environment allows the process to unfold without the water damage. This is just another

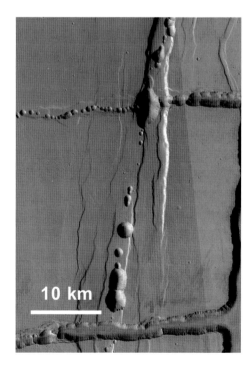

BELOW: Tractus Catena, located just under one thousand kilometers north of Ascraeus Mons, is an impressive series of pits arranged along part of Tractus Fossae, a fracture that is part of the extensive radial array of fractures around the Tharsis Montes.

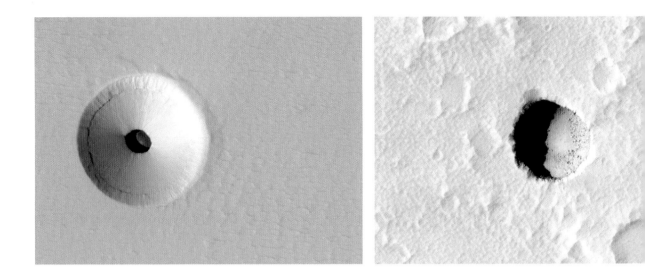

example of why Mars is such a geologically interesting place. Many phenomena that are modified by water on Earth are perfectly preserved on Mars. Hence a geologist who wants to understand a purely tectonic or structural geologic process, without all the erosional damage done to it by water on Earth, will have the entire process preserved for analysis on Mars.

In recent years simple, somewhat related deep pits, called cavi, have been detected in high-resolution images and have caused quite a stir because they appear to be dark holes in the surface of Mars, perhaps opening out into enormous caverns below the surface. Many occur along the sinuous courses of lava flows, along volcanic fissures, or in otherwise featureless volcanic terrains. Some of these are assumed to be similar to the pits, sometimes referred to as "skylights" that often develop over subsurface lava tubes within lava flows. As the roofs collapse over time the surface drops into the lava tube resulting in a sunken area at the surface. Others may be similar collapses over openings left behind along volcanic fissures, somewhat similar to the catena along the floor of the graben.

Pits have been the subject of interest in recent years as possible sites for refuge by future Mars explorers from the harsh radiation environment at the surface. They really do represent ready-to-move-in abodes with attention to a few fixer-upper details. Lava tubes and fissure-related caverns can be enormous spaces that could be used for entire communities once sealed and properly engineered for habitation. The other thing about these that excites the community of Mars scientists searching for

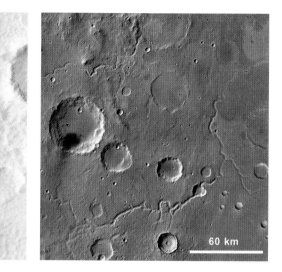

60 km

life on Mars is the fact that they can also represent environments that provide protection and certain mineral nutrients for any past or present microbial life.

Valles. These are a Martian specialty. Although one of the things that makes the geologic characteristics of Mars so fascinating is the near absence, for most of its geologic history, of the damaging effects of water on original volcanic and tectonic features, the presence of water-related landforms has merited some of the most intense interest for a variety of reasons. This has in turn led more recently to the expansion of Martian sedimentology, that is, the study of the sediments and sedimentary rocks of Mars.

Deposits of sedimentary layers and deltas have now been identified near the terminus of many valleys, leading to the realization that in addition to windblown deposits, there is likely a considerable amount of water-deposited sediment on Mars, a concept that embodies the results of many missions and many studies that conclude there was water in Mars's past and that, in many instances, the water transported loose debris and deposited it in low places. In other examples, stacks of sediment layers appear to blanket underlying topography as though they were air fall materials, perhaps deposited as ash layers from one of the past enormous volcanic eruptions. To complicate matters, in some cases the sediments have been stripped away by wind over eons and the underlying terrain that had been buried was being exhumed. These are the types of things that geologists love to decode, like deducing the events at a crime scene from the evidence left behind.

With enough such examples of the evidence from overlapping events, it became a matter of just tracking the sequence of events to figure out a history for an entire area. In the case just mentioned, first there was a terrain, and that terrain was eroded or otherwise modified by geologic processes. Then there was a layer that was deposited on top of that. The intervening surface that was eroded is called an unconformity,

"You have to quit associating beauty with gardens and lawns, you have to get used to an inhuman scale, you have to understand geological time."

—WALLACE STEGNER, "THOUGHTS ON A DRY LAND," 1972

because part of the time is missing between the two events and the overlying rock layers do not "conform," that is, are unconformable with the underlying rocks. Ultimately, when this sort of unraveling of deposits and events is done over broad areas of the planet, a global sequence of events begins to be obvious. At that point it may be possible to correlate similar events at different points on the planet. And finally, all that is left is to compile all these interrelated and correlated events into rocks and surfaces that are similar in relative age and/or similar in rock type. At that point you have a global geologic history.

Unda. The term translates roughly as "undulation" or "water" depending on your Latin preferences, which effectively describes the wavelike and billowing characteristic of dunes. Dunes on Earth are of course a result of wind blowing sand and silt into wavelike ridges and form where abundant sand and little vegetation allows grains of rock and minerals to be concentrated in large fields of deep sand. Conceptually sand dunes are relatively simple. They form when grains of sand are moved along by wind, collect in bulk, and pile up into complex wavelike swell or dunes. And dunes are well known for moving as the sand continually blows off the upwind side and gets moved over the dune crest and piled on the lee side of a dune. As the process continues the dunes eventually "migrate" grain by grain downwind. But while conceptually simple the variety of forms and processes embodied by blowing sand is a whole field of science in itself.

The study of windblown sand and the remarkable array of features related to the process was a well-developed geologic subspecialty on Earth, often referred to as aeolian geomorphology, even before we identified aeolian landforms on Mars. But sand dunes are limited to remote arid areas or small regions where sand is abundant such as beaches or deserts on Earth and their relevance to most societal needs is limited. Hence the opportunities to study them have been relegated to a few academic scientists. Nonetheless the process was well studied because of the relatively simple parameters that controlled their formation and motion, including wind and how it moves particles. But it was an epical study of the process by Ralph Alger Bagnold, *The Physics of Blown Sand and Desert Dunes*, published in 1941 and rediscovered by Mars scientists, that formed the basis for a whole new science of aeolian processes at Mars. Bagnold's work was a subspecialty in the science of geomorphology and was so thorough that for many years following its publication it was the basic reference for aeolian studies and was thus a natural source of all kinds of information once we realized the significance of blowing sand on Mars. Before then there had been little expanded geological research on that topic. An interesting side note here is the little-known fact that Ralph Bagnold's sister was Enid Bagnold, the author of works including the 1935 novel *National Velvet* and the 1955 play *The Chalk Garden*, both of which became popular films. In the words of Sherlock Holmes, commenting on his famous artist relative, "Art in the blood is likely to take the strangest forms."

While the "sands of Mars" was a theme long before the space age, and indeed was the title of the first book published by Arthur C. Clarke in 1951, the identification of the sheer variety of windblown features in the early Mariner 9 and Viking images spawned a whole field of study. The identification of dunes on Mars may be said to have rescued the field of aeolian studies from geomorphological obscurity. We suddenly had a need for understanding how aeolian processes worked in order to understand a problem on another planet. This has happened frequently with other topics since the advent of the space age. While it was pretty obvious that dunelike features on Mars were some sort of windblown form, the geographic nomenclature required avoiding the use of genetic terms, so unda was chosen. In addition to that, a few other questions suggested that the Latin naming theme for this feature was still a good idea. One of those questions was the fact that while we generally refer to these types of landforms as "sand dunes" on Earth, there was a particular problem with using that combination of words for the Martian dunes.

Although to a non-geologist, "sand" usually means the common tan beach sand, it has a very technical definition for the geologist. It can be any material, but it has to be a certain size range. And it was unclear how sand-size particles could be moved in the thin Martian atmosphere regardless of wind velocity. For a long time it was believed that sand grains, which are a specific size of grain larger than silt grains between about 0.06 and 2 millimeters in diameter, could be moved along the surface in high winds only on Earth because the density of Earth's atmosphere and the corresponding force on grains of sand was sufficient to shove each grain along. Based on the fluid dynamics of air moving across sand particles first established by Bagnold's work, in theory the Martian atmosphere is so thin that the force of wind on grains as large as sand was insufficient to move sand-size particles. The Martian atmosphere was only likely to be capable of moving smaller, dust-size particles. Martian

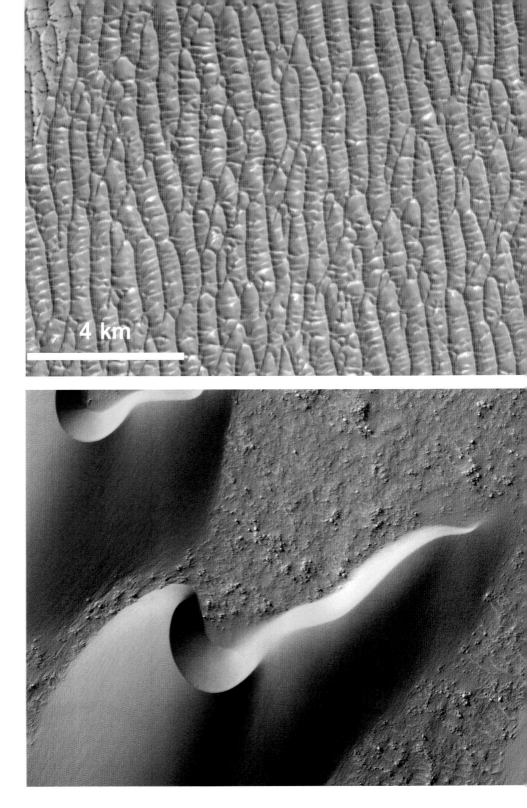

RIGHT: Olympia Undae is part of the vast sand sea surrounding the north polar plateau. Location 81ºN, 180ºE.

4 km

RIGHT: Field of barchan dunes on the floor of Arkhangelsky Crater located at 40.8ºS, 335ºE. This image is about five kilometers across.

"sand dunes" were suspected of being more a type of "dust dunes." This dominated the thinking for decades. But we were proven wrong based on the first look at dune-like ripples on the surface by the rover Spirit where the grains making up the ripples were decidedly sand-size.

Undae on Mars are identified over significant areas of Mars because Mars is obviously a dry planet over which erosion and breakdown of rocks has had ample time to occur. Hence there is a significant source of sand and dust to be mobilized by winds. Undae are to be found almost anywhere on Mars but one of the largest areas occurs in the Vastitas Borealis surrounding the north Martian pole. This is a veritable sand sea rivaling the Sahara in size and is an aeolian scientist's paradise. Abalos Undae, Siton Undae, Olympia Undae (the largest), and Aspledon Undae encircle much of Planum Boreum, the north polar cap. Seen from orbit Olympia Undae is a mesmerizing landscape of dunes of uniform size and orientation that are frequently dusted with frost. Unlike most dunes elsewhere on Mars these are compositionally interesting because they are formed from gypsum, much like the sands of White Sands National Park in New Mexico.

This brings up the question about what the other dunes of Mars are made of. On Earth a close look at the sand making up dunes, or beach sand for that matter, shows that it consists of tiny grains of quartz. There are a number of reasons for this. First quartz is a very stable material and one of the harder minerals and resists the corrosive effects of water, so it sticks around long after other rock-forming mineral grains have been pulverized to dust or water damaged, that is, altered, to clay and other oxides. The second reason that quartz grains are common is the simple fact that much of the rock on Earth from which sediment grains are derived consists of high silica such as granite and various related igneous rocks. Quartz is a dime a bucket on Earth. Right now, I am sitting on a mile or two of the stuff filling a great rift basin here in the center of New Mexico. Most dry areas of Earth are filthy with quartz sand, and we are practically buried in the stuff. But on Mars, free quartz, that is, quartz that is in its crystalline and granular form, is probably pretty rare. It is interesting to speculate that because free quartz is so rare on Mars that a bucket of quartz sand might fetch diamond-like prices on the future Martian mineral market. So what are the minerals of sand grains on Mars if not grains of quartz like that in Earth sand? Remember that sand can be anything so long as it is in that size range between 0.06 millimeters and 2 millimeters. Whereas quartz-rich granitic rocks are common on Earth, on Mars the most common rock is basalt. So most Martian sands are made of small particles of basalt.

> [T]he atmosphere at the surface of Mars is as thin as the air at 45 kilometers on Earth, which is so thin that a space suit would be required to survive the low pressure.

Dunes and ripples occur at many scales, from the giants several meters high in the north polar sand erg to large dunes in the floors of many craters. There are many other manifestations of windblown materials on Mars. Some of the initially fascinating shapes included long trailing ridges downwind from the raised rims of impact craters and long ridges, known as yardangs, sculpted by abrasion from continuous streaming of grains across the surface. Ronald Greeley of Arizona State University was an early investigator of aeolian processes as applied to understanding Martian aeolian landforms. His studies included building large wind tunnels to explore the shapes of wind tails around craters and he even set up a large-scale wind-flow observation tower for understanding wind streaks forming downwind of Amboy Crater, a cinder cone in the Mojave Desert. Whole troops of graduate students worked with Ron on aeolian projects. And this was just the start. Subsequently the field has exploded, with a list of research papers pages and pages long. There is even an annual conference on planetary aeolian processes.

BEYOND NAMES ON A MAP

MARS IS OBVIOUSLY A GEOLOGICALLY COMPLEX planet, and there are many characteristics that you will not find on a map but are nonetheless important things to include in our brief survey, for example, the atmosphere. We know it is thin by Earth standards, about one one-hundredth that of Earth; the atmospheric pressure is about six millibars at the mean elevation of the surface, so the six-millibar level is a kind of Martian sea level substitute. Comparing this pressure with the one thousand millibars of Earth sea level, this means that the atmosphere at the surface of Mars is as thin as the air at forty-five kilometers on Earth, which is so thin that a space suit would be required to survive the low pressure. But then the atmosphere of Mars is also mostly CO_2 as well, a poisonous gas for us Earthlings.

Here is the really interesting thing about the atmosphere: every winter, a significant fraction, up to 30 percent, of the atmosphere condenses out at the poles. This means the atmospheric pressure is a millibar or two lower during one of the polar winters. The reason of course is that the atmosphere is mostly CO_2, and the seasonal expansion of the polar caps is CO_2 ice condensed out of the atmosphere. Fortunately, here on Earth, the polar caps are made of just water ice instead of one of the principal gases that make up our atmosphere, so the air pressure on Earth does not appreciably vary by season.

Another gas in the Martian atmosphere is argon, a relatively inert gas that out-gases from the planet and hangs around. It makes up between 1.5 and 2 percent of the atmosphere and varies seasonally. The seasonal variation is thought to be a result of the simple fact that unlike CO_2, which makes up most of the atmosphere, argon does not condense out during the winters, so it stays in the air and therefore becomes a greater portion of the atmospheric composition when the CO_2 abundance falls in the winter. We did a little experiment on the Mars Exploration Rovers to measure this annual rise and fall of the Martian atmospheric pressure, not by measuring the pressure but by measuring the argon concentration in the atmosphere with the elemental analysis instrument, the Alpha Particle X-ray Spectrometer (APXS). In another example of using an instrument for a measurement that it was not originally designed for, when there were no other important rock measurements going on, we simply let the APXS hang out to the side and measure the argon in the air. Mars Exploration Rover science team member Tom Economou rode herd on the measurements, requesting the activity whenever things were quiet on the APXS schedule. In fact, he was a persistent fellow. Tom is an otherwise quiet grandfatherly type, and we got used to his persistent appearance at daily operations meetings. Just as the team was in the midst of planning and things looked to be about wrapped up, Tom would suddenly appear to remind us that we could be doing another argon measurement. But his persistence paid off and sure enough, the argon fraction of the air was observed to rise and fall with the seasons.

Clouds of Mars are another entire field of study. Clouds have been observed in telescopic observations long before the space age as periodic bright spots and streaks. But spacecraft observations have given a more detailed view. Some appear as orographic clouds forming on the lee side of the enormous volcanoes, others as ground fogs in the canyons. There are even seasonal clouds that can be seen from landers and rovers on the ground. Clouds have been imaged as short movies showing filamentous streaks drifting overhead, often during the early spring when moisture is released from one of the poles. Most of the clouds are actually water ice, somewhat like cirrus clouds common on Earth, especially during winter months. While Mars's atmosphere is incredibly dry, the air is so cold that even the trace amounts are sufficient to generate cirrus-like wisps.

FIVE

New World, New Geology

"Time present and time past
Are both perhaps present in time future
And time future contained in time past."
—T. S. ELIOT, *Four Quartets*

YOU CANNOT DRAW A TREND ON a chart with fewer than three data points. The more data points you have, the closer the trend can be specified. For a long time, planetary exploration didn't provide geologists with more than one data point, Earth. Finally, that was starting to change. Our knowledge of planets was no longer just Earth. As we expanded outward into the solar system we finally started putting more data points on the chart of possible geologies.

With planetary exploration, we were confronted for the first time with a new way of exploring a world and that way was unlike the process that had unfolded in our exploration of Earth. The new process of exploring Mars was, in fact, the complete reverse. With planetary exploration in general, and Mars in particular, we started with some global views, getting a sense of the larger characteristics of the new world of Mars. With each spacecraft sent to Mars, that view sharpened and the details about the greater geography and geology deepened. Eventually we came to collect enough information to begin making some broad generalizations about what processes formed and modified the surface globally and how all the observations were making a kind of sense that provided an insight into the way the new world worked. This global view had given us an advantage in seeing the big picture and then picking out the individual stories to fill in the picture. It was like a giant jigsaw puzzle where the border of the puzzle was already available and all we had to do was find the pieces that filled the inside of that frame.

To understand why this was a uniquely new perspective and very different from the way we had done it on Earth, we can just look back over the long centuries when we were beginning to study and understand geology on Earth. As people began to rationally understand the world around them, they had a few pieces of the geological puzzle that were laboriously put together, in local areas, on the ground, looking at specific rock layers or landscape features. And those were collected as a few unconnected groups of puzzle pieces. But the overall encompassing frame of the puzzle was yet to be found in the box of puzzle pieces. For centuries we were on the ground looking at individual rocks and outcrops and we had no way to get an all-encompassing view from above. Recall the mantra that the best geologist was the geologist who had seen the most rocks. But for a long time, with travel restricted to a few miles from one's birthplace, the process of discovery was a very local exercise. Curious naturalists saw the same type of rock from day to day. When you only see one example and have no context for realizing that rocks are actually very diverse, what else could you say? A rock was just a rock. So the story of Earth was literally hammered out outcrop by outcrop over the years with enough scientists seeing enough rocks and diversity until a mass of geologic information had been acquired to begin making some generalizations and putting a border on the giant picture puzzle of Earth geology.

On Earth each puzzle piece was more or less clearly resolved, but for many years it was the big picture that was absent. By contrast the new science of planetary geology was approaching the problem from the opposite end: we had the big picture, or borders of the picture puzzle, but what were the rocks really like? It is as if each of the puzzle pieces that we could fit together actually had puzzles inside each piece. We had the overall picture, but even some of the details inside the puzzle pieces were not yet resolved enough to provide the full understanding of the big picture. This is where we were in the study of Mars at the end of the nineties.

Mars was a puzzle of many landforms for which we needed more information in order to interpret their true origins and to understand what they were telling us about Mars. But the picture, however poorly resolved at the level of smaller details, nonetheless gave a hint at what the important questions were going to be as we continued the exploration.

GEOLOGIC MAPPING OF MARS

"A map, it is said, organizes wonder."

—ELLEN MELOY,
The Cheater's Waltz: Beauty and
Violence in the Desert Southwest

USING THE GEOLOGIC KNOWLEDGE OF MARS built from the Mariner 9 and Viking missions global images, we eventually had enough information to do a global geologic map of Mars. How this global map came about is a story of work by many researchers over the years dissecting the overlapping relationships between features on the surface. Geologic mapping is, of course, different from geographic mapping, or any other type of mapping for that matter. Everyone is familiar with what a map is. It lays out a view from above showing the principal features of interest in an area such as roads, rivers, hills and mountains, buildings, and anything that might be helpful for the user. Road maps or Google maps are within everyone's experience. But geologic maps go a step further and record the types and ages of rocks and the types of structures within those rocks over an area. It is not only a process of

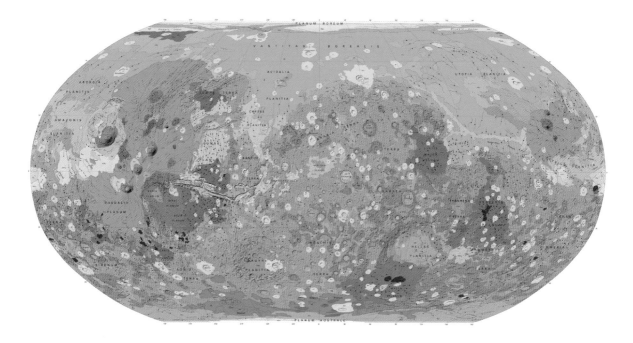

BELOW: Global geologic map of Mars. Different colors represent the variety of geologic surfaces and different ages as determined from analysis of thousands of images by many researchers over many years. This is the latest version released in 2014 of many maps produced over decades. Geologic maps like this one are accompanied by detailed descriptions of each unit and their interpretations based on extensive research analysis.

identifying distinct types of rocks and further identifying their sequence of emplacement. From the overall result a picture emerges of the geologic sequence of events that created the terrain in question. By doing this over an entire planet, the series of things that have happened over geologic time is laid bare to the mind. By doing this for a small area of a planet, the sequence of the events in that small area becomes apparent. It's similar to a type of time travel in which you back out from the available record of rocks what has happened in the past, sometimes the deep past of millions of years ago and sometimes the more recent past.

Geologic mapping, even on Earth, is a type of true exploration, but it is a type of exploration that has not been apparent to the non-geologist. This is because the number of people who have the training and opportunity to geologically map in the field on Earth is vanishingly small; few people have had the experience of the raw discovery that it entails. When you say that you are "mapping an area," the non-geologist will often get a quizzical look and ask, "Does that means you are 'surveying'?" or "Hasn't the area here been surveyed before?"

The concept of geologic mapping is completely beyond the experience of most people. As a result, perhaps it is not widely appreciated that when you are mapping an area, there is a sense of true exploration simply because you are the first person in the history of the world to truly "see" the development and evolution of a landscape, in this case a geologic one. It is a shame that there are not as many amateur geologic mappers as there are amateur rock hounds who collect interesting minerals, rocks, and fossils. The process of unraveling the geology of even a very small area can be a fascinating process and would open up a whole new world of hiking adventure for amateurs who want to get out in the wilds and do something new and interesting. And so, with the geologic mapping of another planet the process holds an even greater sense of discovery and excitement. Imagine, you are the first to explore a whole new world!

There is a catch to doing this with Mars, however. Until recently the only mapping that could be done was what geologists refer to as photogeologic mapping. For several decades our understanding of all the other planets has come from orbital data, including photos—including visible and near-infrared light images—and various other spectral imaging methods. Early in the space age the methods for geologic mapping with photos were established during efforts to understand the geology of the Moon as we began the program to send people there. This photogeologic mapping was spearheaded by amazing and innovative geologists, many of whom were from the US Geological Survey. These included Don Wilhelms, Gene Shoemaker, Jack McCauley (then the chief of the astrogeology branch of the US Geological Survey), Larry Soderblom, David Scott, Mike Carr, and George McGill, among several others. Prominent in that group was Baerbel Lucchitta, one of the first women in the field of astrogeology, at a time when there were, regrettably, few women doing research in planetary geology. The process of planetary photogeologic mapping relied on the overlapping relationships between distinctly different terrains visible in the images. In "Geologic Mapping of the Second Planet, Interagency Report: Astrogeology 55," Don Wilhelms laid out the details of geologic mapping just using photos based on his early studies in preparing for landing site selection on the Moon. He explained that the methods were codified in order to eliminate what he referred to as the many "parageology" studies of the Moon prior to 1960 and sought to introduce a system of more rigorous methods for extracting geologic information from photos of the Moon. At its heart, then, photogeologic mapping is simply the process of identifying distinct rock masses from their characteristics as seen in photos of the surface, and then noting their relative sequence, that is, where one unit overlaps another or where other evidence implies that one unit preceded another in its formation. This is simply the old principle of superposition. Like making a layer cake, the first layer is laid down on a plate and then the second (slightly younger) on top of it, and so forth. Geologists call this the process of stratigraphic sequence in which rock layers are observed and the relative ages of the complete stack of rock or terrain units is then apparent. Wilhelms remarked:

> *The best maps have been produced by experienced field geologists who understand the purpose, strengths, and limitations of geologic maps; who see their utility in lunar and planetary studies even in the absence of final data; who are willing to apply their research methods and understanding of terrestrial geologic relations and processes to other planets; who are patient and careful; and who have no hang-ups about extra-terrestrial bodies. There is a close empirical correlation in quality between a geologist's lunar and terrestrial maps. Geologists who have made at least one complete and good terrestrial map from field studies generally have been able to make good lunar geologic maps, if they wanted to.*

And you might remember that my graduate advisor, Wolfgang Elston, thought the same thing. Therefore, my career as a planetary geologist began with mapping here on Earth.

AN ALIEN GEOLOGIC TIME FOR AN ALIEN GEOLOGY

FROM THE COMPREHENSIVE EXERCISE OF MAPPING the surface geology of Mars, the detailed geologic history or sequence of Mars emerged. The results were a significant accomplishment in deciphering Mars, and certain patterns were recognized that form the basis for many subsequent questions. Things happen slowly on Mars and most of the exciting big surface features are incredibly old. But geology and geologic mapping lets us do virtual time travel. Just ask any kid who has ever visited a natural history museum. Many types of fantastic extinct beasts and exotic environments are revealed by looking at the record of fossils in rocks laid down millions of years ago. By looking at ancient rocks, we can see and learn from the changes in life-forms and environment that have occurred throughout time. From the information extracted from a single rock outcrop we can step back to a time when life on our planet in the strict sense was literally very different and to see where we have come from and guess, of course, at where we are going.

The first orbital missions to Mars allowed us to do on Mars the same kind of global geologic mapping we had done on the Moon, and a version of the type we had done on Earth. Following the Mariner missions, nearly the entire surface of Mars had been geologically mapped and a global synthesis was undertaken and completed by 1978 and published by David Scott and Mike Carr. In assembling the original global geologic map, it was then simply a process of arranging the hundreds of geologic terrain units mapped on the surface of Mars into a global stratigraphy based on the crater abundances on each of these geologic units. Armed with this new tool of planetary mapping, the geologic mapping of Mars began soon after the first Mariner 9 image mosaic quadrangles were assembled and has continued through Viking and all the later missions at ever increasing levels of detail.

The Viking Orbiters gave us the high-resolution images necessary to do that even better than we had done before with Mariner 9 images. This mapping set the stage for an understanding of deep time on Mars. Three distinct times during which certain types of geologic landforms were common emerged. This resulted in the first global stratigraphy of Mars in which three eras in Mars geologic history were initially defined based on a combination of relative ages and distinctive things about each successive age. The resulting three eras of Martian geologic time were given the names Noachian Eon (oldest), Hesperian Eon (intermediate), and Amazonian Eon (youngest).

The global mapping was further refined with the more detailed information following the Viking mission. In an epochal compilation of many individual geologic maps, Ken Tanaka, David Scott, John Guest, and Ron Greeley completed geologic maps of the western and eastern hemispheres of Mars. There have been several iterations and tweaks to the original map, including conversion to a digital geographic information system (GIS) format. But the results are still impressive and informative. More important, Tanaka, who was the US Geological Survey planetary mapping lead at the Flagstaff Astrogeology Science Center and had overseen all of this, published a summary of the Martian stratigraphy in 1986 and laid out the best estimates for the possible ages of these Martian geologic time periods.

Of course, we did not have actual dates or so-called absolute ages for these Martian geologic time periods, but we did have another more or less quantitative estimate of the relative difference in ages by determining the density of craters on a geologic unit. Crater counting, as we will see shortly, has been one of the pillars of planetary geologic mapping since the early days of lunar exploration and relies on determining the number of craters accumulated by any geologic surface in a photogeologic map. The more craters, the longer the surface has been exposed to things that impact a surface and make craters. Think of it in the same way as cracks and potholes on a city street. If you drive down one street with many cracks and potholes and then turn onto another street with no cracks or potholes, you know instantly which street was recently repaved. To put it simply, the older the surface the more craters that have accumulated on it.

But older and younger are relative terms. How much older? Exactly how young? What are the true ages of the surfaces associated with these Martian geologic eons? There are a couple of ways of estimating the ages of a surface associated with accumulated numbers of craters. We had some experience with assigning actual ages to geologic units on the Moon. There we had some actual dates on samples returned by the Apollo mission. Those age determinations use the same radiometric methods used on Earth to determine the age of rocks. Then it was a relatively straightforward process of correlating the determined absolute ages from dating the rocks with observed abundances of craters associated with the unit the rocks came from. But on Mars, where we have not yet collected and returned samples of particular geologic units, it is necessary to make some estimates based on assumed rate of crater formation. As with any estimation process, it is an approximation. But the science behind the approximation is strong, so the estimates can be supplemented with actual dates once we start analyzing samples of Mars material here on Earth someday.

Where did we get these names for the different eons of Martian geologic time? It so happens that the Noachian Eon was the period of great deluges of water and coursing rivers that flooded the Martian landscape. The name Noachian conjures up the thought that it must be from the biblical story of Noah and the Ark in which Noah built a great boat so that he and his family along with a sampling of creatures of every kind could float in safety during the Great Flood that was soon to follow.

> The first orbital missions to Mars allowed us to do the same kind of global geologic mapping we had done on the Moon.

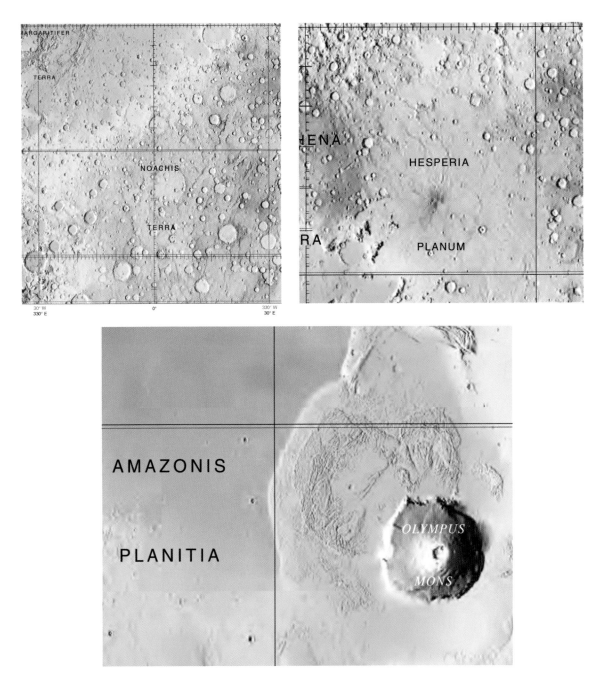

TOP LEFT: Color-coded relief map of Noachis Terra, the type area of the Noachian Eon in the Martian geologic time scale. **TOP RIGHT:** Color-coded relief map of Hesperia Planum, the type area of the Hesperian Eon in the Martian geologic time scale. **ABOVE:** Color-coded relief map of Amazonis Planitia, the type area of the Amazonian Eon in the Martian geologic time scale.

But it is actually a bit more involved than that. The names of geologic eras on Mars were selected from the local geographic names of "type areas" where rocks of a particular age are first described or where the best and most obvious examples of rocks of that geologic age are located. This is common practice in geology in which a place that is representative of something that is common over much larger areas or where the rocks from that time were first defined is referred to as the type area and the name of the type area is used to refer to the specific type of rock or the time in which it formed. Now the area on Mars where geologic map units of the oldest age based on crater abundances, estimated at between 3.7 and 4.1 billion years old, are best exemplified is a region on Mars's cratered southern highlands known as Noachis Terra. It so happens that this is a name that was used in Schiaparelli's original maps of Mars.

Recall that Schiaparelli chose names for his map using classical names and these names were later used for geographic names in modern maps derived from spacecraft images.

Here is the slightly ironic part. Noachis means "land of Noah." So even though there was no knowledge of the actual surface terrain in Schiaparelli's time, it just happened that he used a name for that region of Mars that was in a way entirely appropriate for a terrain that had experienced some of the greatest effects of past water on Mars. Coyote Mars strikes again with a little sleight of hand. On the other hand, it is entirely possible that the name was selected for its association. I do not recall that there is a discussion in the formal scientific definition of these stratigraphic names regarding the thinking behind the naming process. Perhaps it was buried in the original stratigraphic naming committee comments and I missed it, and if so, I will likely be corrected. But the association may have helped. There were several similar highlands with other names to choose from. Such are the methods of exploration and explorers' prerogatives.

The Hesperian Eon is named after a type area on Mars known as Hesperian Planum. This region is characterized by, of course, slightly fewer impact craters than Noachis Terra from which the Noachian Eon got its name. The Hesperian Eon is also characterized by a different type of geology. Instead of the great areas of water-carved channels formed in the Noachian Eon, the Hesperia Eon is named after the Hesperia Planum region, an area of vast plains of basaltic lava flows and associated landforms that are prevalent in those volcanic plains. So the Hesperian Eon is a somewhat more recent time on Mars when volcanism was more common and water-carved channels were less widespread. In fact, many of the lava flows from this time occupy the channels cut by water in the earlier eon.

The Amazonian Eon is named after the type area on Mars known as Amazonis Planitia. This area has fewer craters than the Hesperia Planum region. It is a vast plain dominated by windblown sands and the influences of the cold and dry, relatively dead environment of late Martian geologic history.

None of this is to say that other exciting things did not occur outside the geologic times when they were more typical. We know now that great lava flows have occurred within the past tens of millions of years as did many of the events associated with the great Tharsis volcanoes. And there is suspicious evidence for some water-related features in equivalent times. For this reason, the description of the Amazonian as being a "dead" time is not entirely accurate. Like all complex things, Mars is a bit more complicated than our simple categorizations. And therein lies the interesting work that remains to be done.

But the story of Martian geologic time and stratigraphy is not over yet. There is another way to define stratigraphic time on Mars other than the one that uses the

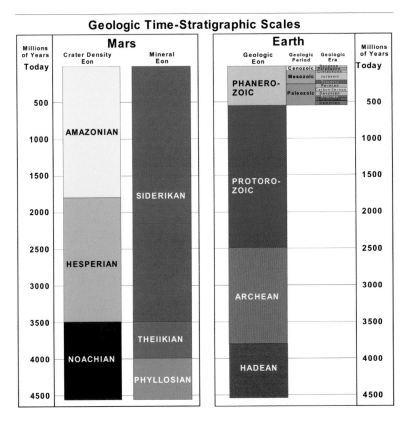

RIGHT: On the left are Martian geologic time scales, one based on the relative ages of surfaces and one based on the times during which certain types of minerals were forming. The scale on the right is the geologic time scale for Earth.

density of craters and the interpreted age of different surfaces. This alternative Martian geologic time scale developed as we began to analyze orbital remote sensing with instruments on the spacecraft that could detect specific minerals. As our knowledge of the true nature of the rocks that represented these different crater ages of the surface became clearer with these more powerful methods of remote sensing, the mineral characteristics became known and another attempt to categorize the Martian time came about. This was a rock or lithologic time series, but defined in terms of the predominant mineral that seemed to be formed during specific periods of Martian geologic history due to the changing climate on Mars with time.

In their attempts to piece together the puzzle of surface compositions that were emerging from later missions including the European Space Agency (ESA) Mars Express and the US Mars Reconnaissance Orbiter, French planetary geologist Jean-Pierre Bibring and colleagues proposed three eras: an early era represented by rocks bearing a suite of minerals called phyllosilicates; an era in which acidic aqueous waters were more common, resulting in rocks rich in sulfates; and an era in which water was largely absent, identified by rocks with a prevalence of ferric (iron) oxides. These new time periods or eons were named the Phyllosian Eon for the time of clays or phyllosilicates, the Theiikian Eon for the time of sulfates, and Siderikan Eon for the time when oxidizing iron was the most common process. Together, these two systems are the two methods, crater density and mineral, similar to the way geologic time periods were originally defined on Earth, from differences in the record of the rocks themselves and differences in the apparent ages of the rocks.

How do these two ways of organizing Martian geologic time with the geologic time scale compare with the geologic time scale on Earth? Perhaps many people are familiar with the names of geologic periods such as Jurassic or Devonian and many of the other names for distinct times and environments defined from the fossil record. But those and other familiar names of geologic time periods are actually small subsets of geologic time on Earth. And they are relatively recent, referring to environments and times within the last six hundred million years when life emerged on dry land for the most part.

The entire time on Earth that is encompassed by the emergence of life on land to the present is known as the Phanerozoic Eon. This is the time period that most of us hear about when the story of geologic time is discussed. But it is only the last small part of geologic time. Earlier than that is the vast time from 600 million years ago to nearly 2.5 billion years ago known as the Proterozoic Eon, in reference to the time when early life-forms, proterozoa, were dominant. And before that is the Archean

Eon from 2.5 billion years to 3.7 billion years, the time of the earliest evidence of life. And even older than that is a time, the Hadean, when Earth was a violent place in which the surface was subject to all sorts of epic formative processes, including frequent giant impacts and volcanism.

The eons of geologic time on Earth are more comparable in timing with the three great eons of Martian geologic time from the two methods of organizing Martian geologic time. This means that the latest eons on Mars are way longer than the Phanerozoic on Earth and extend back in time as far as several billion years like the Proterozoic and Archean Eons of Earth. Things happen slowly on Mars, and most of the exciting big surface features are incredibly old.

Just as interesting is a quick look at the comparison of the crater density eons and the mineral eons of Mars. It so happens that the time of mostly iron oxidizing characterized by the Siderikan (mineral) Eon extends pretty much back to the beginning of the Hesperian (crater density) Eon. And the time of the sulfates and phyllosilicates, the Theiikian and Phyllosian Eons, are pretty much packed into the time of all the great water events, the Noachian (crater density) Eon. You can look at Mars geologic time from at least two different perspectives of age and environments, yet the picture that emerges is one in which many very ancient events are fairly well-preserved in the modern surface in one form or another. The older events are greatly modified, and in some cases entirely erased by geologic events. But in comparison with Earth, the surface of Mars shows a much older record. For this reason, Mars is a good place to see how the history of a planet unfolds with great time, unlike Earth where the geologic processes are so dynamic that many of its early formative events have been either greatly modified or totally erased by its dynamic erosion and resurfacing due to active volcanism and plate tectonics.

We now had a global geologic history, a global map of all the different types of geologic features and presumably rock types, and an idea of the relative ages of those different features and terrains. We had developed what is known as a global stratigraphy for Mars. To summarize, this told us that regardless of how you named them or defined the surface ages, some of the wildest and wettest events recorded in the rocks making up the surface of Mars were among the oldest on Mars.

Very quickly we realized that many of these things would be much better understood if we could only go down to the surface and stand on it. If we could just get down and hammer on some of those rocks the way we did to unravel Earth's history, who knows what we might learn. For this reason, the story of understanding Mars has been the complete inverse of the way we did it on Earth. On Mars we started with the big picture and drilled down to the details, whereas on Earth we had the more confusing task of starting with the outcrops and building a big picture. Both methods work, but, for this geologist, the Mars way has been more fun.

SOLDIERING ON IN A MARS EXPLORATION HIATUS

THE VIKING PROJECT OF THE LATE 1970s was an epic game changer for Mars exploration. We had nailed down the principle events that shaped the surface of Mars. It was a good thing that it happened using Viking data too, because few additional Mars missions took place for the next twenty years or so. Mars launch opportunities occur on a twenty-six-month cycle due to the need to take the shortest route to Mars, which translates to smaller rockets. But to add to the misery of this long exploration dry spell, some first missions that were sent late in that twenty-year time frame were unsuccessful. Before the next successful mission, this post-Viking mission view of Mars was what we had to work with over the next couple of decades. It was these data that we used to formulate many of the questions that guided our future endeavors in the quest for an understanding of Mars. Although we had compiled the main menu of Mars, we did not as yet have a very enlightened understanding of the detailed ingredients. We knew a lot of the *what*, but we were still working on many of the *how*s.

BELOW: Successful Mars spacecraft missions over the last eight decades. Many of us refer to the gap in exploration between Viking and Mars Global Surveyor, the late seventies to the midnineties, as the lost generation.

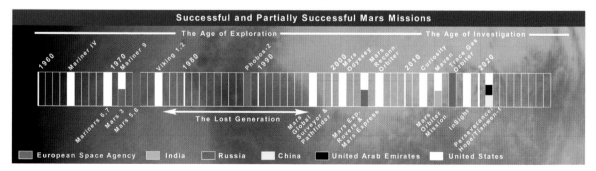

The consequences of this long dry period in Mars exploration by NASA is not often talked about, but were significant for the second generation of planetary geologists, that is, those like me, who came following the first wave of Mars scientists during the first missions to Mars in the late sixties and early seventies. After the heady days of the sixties and seventies, when we were sending spacecraft to Mars every two years and the exploration of Mars appeared to be on an exciting trajectory of ever-increasing new discoveries, suddenly, new missions to Mars completely stopped. That meant no funding and very few opportunities available for Mars geologists.

Little did we know at the time that it would be twenty-one years before another spacecraft landed successfully on Mars. Much the same thing happened with missions to the Moon. Following the Apollo program's completion with the Apollo 17 mission in late 1972, nothing new was happening on the Moon. In the case of Mars, after Viking, NASA moved on to other projects, particularly the outer planets of Jupiter, Saturn, Uranus, and Neptune and inner planets Venus and Mercury. Certainly, for the Moon, there was an abiding sense of "been there-done that" or "mission accomplished." Whether that was true or not, most of NASA's resources appeared to be tied up in the space shuttle and a series of missions to the outer and other planets.

"And sure enough, even waiting will end . . . if you can just wait long enough."

—WILLIAM FAULKNER

But the consequences were significant for those of us in our early and, increasingly so, mid-careers studying Mars. Essentially, the only planetary geology data for Mars during the eighties and early nineties were still the data from Viking. And we squeezed those data for every drop of solid information, and then some. Mars was mapped geologically and remapped at ever more detailed scales with Viking data. Every characteristic of its geology was spun off as a subject for intense specialty study in many research papers. It was no longer an easy task for one person to follow the whole of Mars research. You were a crater mapper, a crater counter, a windblown feature studier, a Mars climate researcher, a Mars volcanology guru, a multispectral Mars image analyzer, or even a Mars geochemistry master. We even had rocks, certain meteorites, for which there was good evidence of an origin on Mars. So there were those who studied Martian rocks in the lab.

All of this research set the stage for missions that were ultimately to follow, beginning finally in the late nineties and continuing to the present. Armed with a global geologic map we could also now make educated attempts to narrow the research to those areas where the most science could be had with later missions. The results of the Mariner and Viking missions therefore provided a template for investigation of what was to come.

SIX

The Invasion of Mars, Phase 2

"The answers are all out there, we just need
to ask the right questions."
—OSCAR WILDE

EXPLORATION UP TO THE MIDDLE OF the 1990s had completed a reconnaissance in which we had learned the most fundamental details of Mars's surface as seen from above. But many questions remained about the fundamental origin and significance of what we had seen up to this point. The only way to answer those questions was to continue pushing onward into the unexplored "interior" of this new landscape with better and higher-resolution orbital instruments. We did not as yet have a good sense of the geology at the finer scale that would permit understanding how these features came about. Nor did we have answers to fundamental questions such as the history of water on Mars. With Viking, we had developed a whole array of specialty topics such as wind features, volcanism, and water-carved landscapes. These specialties set the stage for an ever-increasing array of Mars questions for which answers could only be had with more data from new missions.

MARS FIRES THE FIRST SALVO AT EARTH: MARTIAN METEORITES

AFTER VIKING, THERE WERE NO NEW successful missions to Mars for nearly two decades and new things in the world of Mars science were few and far between. The eighties became a time of data analysis, geologic mapping, and sorting out many of the interesting observations that had been made by Mariner 9 and Viking. However, there was a much-needed jolt of Mars research adrenaline and excitement when it was determined, during the early eighties, that Mars had in a way already jumped the gun and had in fact invaded us instead. Mars it seems had sent rocks smashing to Earth over the years. It is a long story. The study of meteorites is an extensively developed field of science. There are many distinct types of meteorites besides the usual iron meteorites that many of us space scientists are constantly identifying for the public, or more commonly, identifying magnetic and metallic rocks and other objects as "not a meteorite." Actually, many meteorites are nonmetallic, or what are called "stony," and look very similar to Earth rocks in composition. Almost all "ordinary" meteorites are very old, usually dating from 4.6 billion years ago, and hence likely fragments of bodies like asteroids that pretty much formed and cooled in the earliest days of the solar system. In fact, this very ancient origin is why most meteorites are valuable for research; they represent a way of looking at the conditions and the materials present during planet formation. It is very much like geology in general, that is, looking at old rocks tells you about conditions when the old rocks formed. The oldest rocks, meteorites, tell us about the earliest times in the solar system.

ABOVE: The Martian meteorite "Black Beauty," or NWA 7034, is an usual sample containing traces of Martian water.

The science that studies meteorites is known as meteoritics and the scientists, called meteoriticists, have developed some very sophisticated analytical tools for extracting information from minerals at the microscopic level. The same tools were used for study of the returned lunar samples and other "astro" materials.

Among the stony meteorites is a strange class that is not so old and that is curious in itself. For many years, some have been known to be only approximately 150 to 575 million years old, which by meteorite standards is incredibly young, getting up there with many of the rocks on Earth, and certainly a far cry from the billions of years typical for most meteorites. Not only that, these young meteorites are essentially volcanic, whereas most of the older iron and ordinary stony meteorites are not.

This was of course an enigma, because volcanic rocks only happen on volcanically active planets, and stranger still, all the asteroids have very old surfaces where the volcanism that did occur was likely billions of years ago mostly at the same time the solar system was forming. The only two places with possible volcanism that young would be Venus and Mars. Early in the process of categorizing these unusual meteorites it was recognized that there were three distinct types represented by the meteorites named Shergotty, Nakhla, and Chassigny. Meteorites are named after nearby localities where they are found. Shergotty fell at Sherghati, India, in 1865. Nakhla fell in El Nakhla, Alexandria, Egypt, in 1911. And Chassigny fell at Chassigny, Haute-Marne, France, in 1815. These three distinctly different types of odd stony meteorites, with unusually young ages—evidence of some alteration—and distinctive isotopic compositions of included gases, became the basis for a group known as the SNCs, sometimes informally referred to verbally as the "snicks."

The identification of the SNCs as having a Martian origin is a modern Sherlock Holmes detective story and illustrates something about how science works. The unusual class of meteorites had been known for some time by the early eighties, and several researchers by this time had proposed that many of the characteristics were suspiciously Martian. The composition of Shergotty in particular was similar to rocks that the Viking Landers had analyzed. On top of that, the details of the geochemical composition were odd. In addition, the isotopic composition of gases trapped in the meteorite was similar to the atmosphere observed on Mars by the Viking Landers. And then there was the fact that these meteorites showed evidence of alteration by water, something not prevalent in other meteorites. By the time enough studies had analyzed and compared results in the mideighties, it was difficult to conclude anything else but that these rocks were from Mars. If not, there was another planet out there that was Mars's twin.

BELOW: Part of a display of Martian meteorites on loan from the University of New Mexico Institute of Meteoritics in an exhibit in 2015 at the New Mexico Museum of Natural History and Science. Martian meteorites are an exciting way to introduce the public to how science works. In this case it was presented as a modern detective story.

Less than one-half of 1 percent of the known sixty-one thousand meteorites have been identified as Martian meteorites, so they are very rare. Some of them were identified long after they had fallen or had been found. There is one famous Martian meteorite that had been sitting in the backyard shed of a homeowner in Los Angeles for years before it was recognized for what it was. But many Martian meteorites had been found over the previous few years in the homogeneous terrain of the North African desert sands or during annual collecting trips by meteoriticists into the snowy interior of Antarctica, where rocks sitting on the surface look a bit out of place. By the way, lunar meteorites have also been identified. But what about the other planets? The key to identifying the Martian and lunar meteorites was of course based on the simple fact that we had collected information about Martian gases on the Viking missions and actual lunar samples from Apollo. So, for those two bodies we had something to compare the meteorites with and identify the necessary geochemical "fingerprints." But if we, in fact, have meteorites from other planets, which would seem to be reasonable, we cannot confirm it because we do not have actual information about the surface properties from any samples on those planets because we have not visited the surfaces of those planets or returned samples from them. Who knows, maybe somewhere there is a meteorite from Venus that we just can't recognize or identify as being from that amazing planet. It is an interesting speculation, but we just cannot confirm or deny it because we simply do not yet know the necessary information about the surfaces or atmospheres of other planets.

The story of Martian meteorites received a burst of celebrity in 1996 when it was announced that one meteorite with the unassuming name ALH 84001 (meaning that it was found at Allan Hills in Antarctica in 1984) might preserve evidence of past life on Mars. The late nineties were already a time of revived activity in the exploration of Mars. But the curious story of Martian meteorites was getting in on the act. The story is well chronicled, but in short, in an article published in *Science* authored by David McKay, Johnson Space Center, and several others, strange microscopic features were reported that looked like small bacteria. The article was a media sensation, and a formal NASA press conference was held. I recall that many of us listening in on the press conference there in the lecture hall at Brown University, where I was by that time a postdoctoral researcher, were a little dubious when we saw the images of the proposed fossil life-forms.

They looked too good to be true. This was not just the usual critical commentary that scientists engage in with anything that is new and very radical. It just had the feel of something that needed a little more work to determine if it was a real possibility. Subsequently other researchers found many difficulties with the interpretation and ultimately it was concluded, for now, that the putative life-forms were more likely nonorganic or the results of things that happened after the meteorite arrived at Earth. But like the initial Viking life experiments in 1976, the story retains some advocates and lives on in some speculations.

Meteorites are a popular field, and given the fact that many finds and observed falls have happened throughout history, a certain amount of lore and strange stories and myths have arisen, especially in times and places where modern science was still in its infancy. There is the curious story of the meteorite Nakhla, later determined as a unique type of Martian meteorite, that as with most sensational stories, appears less so or suspect when examined in detail. The story goes that in the year 1911, on June 28, near the Egyptian village of El Nakhla El Bahariya, there occurred a series of loud booms and streaks across the sky followed by stones raining down on the landscape. As reported by the Arabic newspaper *Al Ahali*, one local resident in the nearby village of Denshal said one of the stones struck a dog "leaving it like ashes in the moment." The story was never actually confirmed that a dog was struck by the meteorite, nor, of course, is it likely that a just-fallen meteorite, which are invariably cool to the touch despite folklore that says otherwise, could have burned an unfortunate dog to ashes.

There are other problems with the story when researched, such as the wrong date and the fact that Denshal is some thirty-five kilometers away from the area of the known fall area, among other things. Nonetheless the story remains as a popular myth that you will hear frequently. The meteorite Nakhla probably did not strike a dog, but it actually turns out to be a rare example of a Martian meteorite. Study has determined that it was formed about 1.5 billion years ago, a major clue to its Martian origin, along with other evidence of a more detailed geochemical nature.

That is a very short synopsis of the Martian meteorite story and a peek into the world of meteoritics. The main point to gather from this is that it was the big story about Mars during the dry spell immediately following Viking. Another point to take away is the fact that in a sense we have Mars samples already here on Earth. So people often ask why we need to go to Mars at all and certainly why we need to bring back samples from there. But there is a big problem with these Martian meteorites as samples. Despite all of the things we can learn about the rocks in a laboratory, we do not know their context. We do not know where they came from on Mars. So we can only speculate which geologic unit that we have mapped that they represent. It is frustrating. But if we could go to Mars and collect a sample now, then we would have something to work with. We would not only know where it came from but also how old that unit is (something we can learn only with laboratory instruments here on Earth), and how it relates then to other adjacent terrain units on Mars. At this point there starts the long quest for what is known as a Mars sample-return mission. We are a ways from that happening, but we are getting closer and closer to making it a reality. In the end, the early days of post-Viking Mars research were not entirely dry, but missions to Mars were just not happening.

Of course, this may not have been the first salvo fired by Mars at Earth. As we had speculated at the beginning of chapter 1, maybe life started on Mars and seeded Earth with a few well-placed Martian-microbe-infested meteorites way back when. Or maybe any emerging Earth life never stood a chance against the alien invaders from Mars, and Mars has the last laugh when in the end we discover that we were "compromised" from the very beginning. But these are the sort of speculations best left for science fiction writers.

"Only he who attempts the absurd is capable of achieving the impossible."

—MIGUEL DE UNAMUNO, *Essays and Soliloquies*

EARTH FIRES A FEW SALVOS BACK AT MARS

AFTER SEVERAL YEARS HAD PASSED SINCE the successes of the Viking mission to Mars and no new missions had been planned or launched, it was beginning to seem like Mars exploration was a thing of the past. During this time I had completed my degree in geology at the University of New Mexico and a PhD in planetary science at the University of Arizona, and both Jayne and I were working at Brown University as planetary geologists with Jim Head, who worked on an amazing range of terrestrial and planetary geology topics, on many planets, and with colleagues from all over the world. We were both fortunate to work with Jim. Tall and very energetic, Jim favored jeans and boots, and it was not unknown for him to be working in his office on Christmas Day, surrounded by his amazing collection of unusual beer bottles arrayed along the shelves full of books. Because of Jim, Jayne and I were also incredibly privileged to work with English, French, Finnish, German, and Japanese planetary scientists, as well as the Russian planetary scientists at the Vernadsky Institute in Moscow, through a cooperative agreement that Jim had arranged between the Brown University Planetary Geosciences Group and the Vernadsky Institute, mostly in association with efforts to get prepared for the coming NASA Magellan mission to Venus. The Russians had considerable early success with sending missions to Venus and the Venera orbiter radar surface mapping missions had provided some early examples of what the Magellan mission would be up against. So Mars was for the moment a thing that was on the table while we explored another planet. But just when things looked bleak for more Mars research, the exploration sputtered to a start and would go on to gradually ramp up to a true return to Mars. Because we were working closely with colleagues in Moscow, we had a somewhat inside view of their programs.

Russian Phobos 1 and Phobos 2. Russian Phobos 1 and 2 were first out of the gate in the post-Viking mission return to Mars in 1988. Both missions were primarily intended to explore the moons of Mars, Phobos and Deimos. The moons of Mars are two diminutive bodies, Phobos being only twenty kilometers in diameter and Deimos somewhat smaller at ten kilometers. As imaged first by Viking, Phobos was found to have an odd potato shape.

Phobos and Deimos orbit Mars fairly closely, taking only seven and half hours and ten and a half hours to orbit Mars, respectively. As a result, as seen frequently

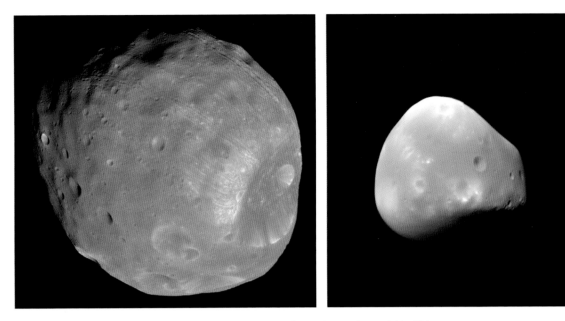

ABOVE: Phobos (left) and Deimos (right), moons of Mars as seen by the Mars Reconnaissance Orbiter High Resolution Imaging Science Experiment (HiRISE) camera in orbit about Mars.

from the surface of Mars by later rovers, each moon moves across the night sky of Mars rapidly, much the way satellites move across the night sky of Earth and not at all like the stately progress of Earth's Moon across the sky in its twenty-eight-day orbit. Because of their small dimensions, they are much more similar to the asteroids that spacecraft have visited in recent years, and are interpreted to be captured asteroids, and thus offer an opportunity to be visited a bit more easily with their very low gravity than the Moon with its stronger gravity. So a spacecraft could cozy up to one of the moons of Mars and even sample the surface with far less technical difficulty. This was part of the goals of the Phobos missions.

Phobos 1 launched July 7, 1988, but had a problem en route due to human error. The problem arose with a command that was sent from Earth while the spacecraft was only a few weeks out. During normal interplanetary cruise phase activities, a command was sent that inadvertently told the spacecraft to turn off the altitude control thrusters. This in turn meant that the spacecraft was unable to orient its solar panels toward the Sun eventually resulting in depleted batteries before the error was identified. The command was an error traced to a small piece of computer code in which a hyphen was left out; that was a command that was purposely in the code for activities on the ground before launch but not intended for flight.

There are other elements to the story, including why the command was still in the software on the spacecraft at all and the failure of some procedures for double-checking commands. But the results were fatal for the spacecraft. There was an investigation to determine who was responsible. But it was decided that any punitive actions would be delayed because Phobos 2 was still ongoing and it was likely to increase the stress on the mission crew. It was reported that IKI (Institut Kosmicheskih Issledovanyi, or Space Research Institute) director Roald Sagdeev made the somewhat callous joke, as viewed in the context of an authoritarian culture, that quoted a former secret service chief under Stalin, "Let's make them work for now. We can shoot them all later."

Of course, you would have to know Sagdeev, who was a very calm, unexcitable, and somewhat soft-spoken senior scientist, to realize the joke was the standard Russian black humor born of years of working within (and sometimes around) the Soviet regime. Jayne and I got to know Sagdeev and many of the other excellent Russian planetary geologists well, and as friends. We were fortunate enough to work with them, both during work visits to the USSR and later, when they were allowed to travel to the United States, at Brown.

Phobos 2 carried a lander with an unusual hopping mechanism with which it was to move on the surface of the Martian moon Phobos. Phobos 2 launched on July 12, 1988, and successfully made its way to Mars, arriving in January 1989. During its orbits of Mars, it would work its way to a rendezvous with the moon Phobos, where it would deploy the hopping roving, a lander with a variety of instruments for measuring the composition of its surface, and a penetrator (to penetrate into the surface) with other instruments designed to measure physical properties of the rocks. Phobos 2 already had some difficulties with its computers by this time that were threatening to compromise the mission.

Finally, on approach to Phobos, after relaying a couple dozen new images of that moon, contact was lost with the spacecraft, later determined to again result from a failed computer. It was excruciatingly bad luck that things went wrong just as the spacecraft was approaching its big event. Also, in the minds of the paranoid, it appeared very unusual given the timing right when the spacecraft was about to approach the moon Phobos. Perhaps it was an echo of the early speculation that has suggested the moons were hollow because of their predicted decaying orbits. This was back in the days before Viking, and I recall reading this strange theory in the newspaper as a kid. But the bad luck in this case with the Phobos 2 mission was immediately alighted upon by the world of UFOs and mysterious conspiracies and ignited speculations that the spacecraft had run afoul of an alien base somewhere on Phobos. And as proof, all manner of odd things were reported by alien enthusiasts as visible in the images, as those who know little about planetary surfaces or imaging systems have been wont to do. The Russians had seen *something,* and they weren't telling the truth! Or so the speculators proposed.

Even the images returned of the Mars surface fell under the review of conspiracy theorists. I recall that during interactions with Soviet colleagues visiting Brown University at the time I was handed one of the better images of the surface taken by the infrared spectrometer TERMOSKAN, which showed for the first time in a decade a new image of Arsia Mons and its equatorial surroundings. I did some photogeologic mapping with this data to see if anything new could be extracted regarding the geology of Arsia Mons, but aside from a beautiful view of the caldera and the fan of lava flows on the volcano's south flank, the resolution was insufficient to really provide significant new insights beyond the Viking data and the research that Jayne and I had done as Viking Student Interns.

But interestingly, within this image was a long, dark shadow spread out in alignment with the equator that was a curious result of the shadow of Phobos on the surface. It happens that the TERMOSKAN instrument acquired the spectral data associated with the imaged area by long "stares" at the surface, and during this time the shadow of Phobos passed across the surface being spread out like you would expect for a moving object, in this case the moving shadow of Phobos, in a time exposure. But conspiracy theorists saw this image, without reference to the actual science content, as a large cigar-shaped object kilometers long on the surface of Mars! This was not the first time that armchair internet image interpreters have suggested otherworldly mysteries and claimed that space scientists are perpetually covering things up. It goes with the territory unfortunately.

US Mars Observer. On the morning of August 21, 1993, three days before the arrival of Mars Observer at Mars, Jayne and I were driving to Brown University with expectations of some wonderful news about the final approach and arrival of Mars Observer at Mars, when a water pump drive belt broke on our car near downtown Providence just short of arrival at the university. We limped into a gas station and parked it until we could do something later, walking the last couple of blocks to the office in the Lincoln Field Building, where the planetary geology group was located on campus. I was in a hurry because I wanted to be there to get in on the news of the successful preparations for final arrival of Mars Observer at Mars. But it turns out that I should have just stayed with the car and got that sorted out because the water pump belt was not the only thing that broke that day. Contact was lost with Mars Observer as it was executing a course-correction maneuver.

Later investigation determined that the likely cause was leakage of fuel and oxidizer into the fuel lines during interplanetary cruise and a subsequent explosion when the engines were restarted to prepare for the course correction during final approach to Mars. This was among the first of the "faster, better, cheaper" mission design efforts that were instituted at the time to reduce cost and permit more frequent missions by using existing spacecraft hardware designs where possible. But the design of the engine was derived from Earth orbital satellites that were not supposed to be dormant for the long periods required for an interplanetary cruise to Mars. Lesson learned.

Mars Observer had launched September 25, 1992, and was to be the first return by the United States to Mars since Viking nearly sixteen years before. The loss of the spacecraft was not an auspicious start for a NASA return to Mars. I recall that none of us were familiar with the failure of a planetary mission—failure of a US Mars spacecraft had not happened since Mariner 8 took a dive into the Atlantic Ocean in the early seventies. So the immediate question that came to mind was "Wait a minute, this means there is no mission . . . this means that the funding for the science team is now off the table."

> "Wait a minute, this means there is no mission . . . this means that the funding for the science team is now off the table."

Researchers live and die by the availability of funding and getting on a mission was a good funding opportunity. And funding opportunities for planetary geology were hard to get. So the sudden demise of a mission before it even started was a culture shock. The good news in my case was that I was not a part of the Mars Observer science team, so I was not seriously impacted directly. But the bad news was that there would be no new data for any of us to work with now and many of us were getting older waiting for a new mission.

US Mars Global Surveyor. Four more years would pass before the next attempt was launched in November 7, 1996, in the form of Mars Global Surveyor. It arrived at Mars on September 11, 1997, and went on to provide some revolutionary new information about Mars, lasting until late 2006. Finally, we were back at Mars. Mars Global Surveyor's success was a result of some important new types of instruments, including the Mars Orbiter Laser Altimeter (MOLA), which completely mapped the

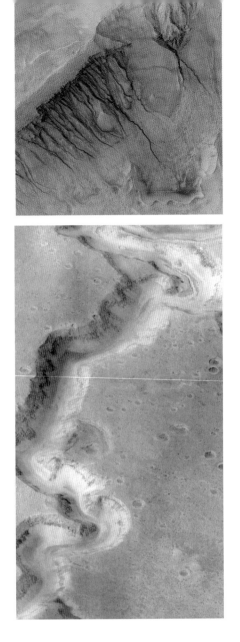

TOP: One of the initial images from Mars Global Surveyor of small gullies on the inner wall of a crater. This became a major topic of debate and investigation in later missions.

ABOVE: Famous image of Nanedi Vallis in Xanthe Terra as seen by the Mars Orbiter Camera aboard the Mars Global Surveyor spacecraft. The valley is 2.5 kilometers wide at the rim. On the floor of the valley is a smaller two-hundred-meter-wide channel.

surface relief or topography for the first time; a Thermal Emission Spectrometer (TES), which provided the first good determinations of mineral compositions of the Martian surface, including discovery of concentrations of hematite later to be prominent in the selection of the landing site for the Opportunity rover; and a Mars Orbiter Camera (MOC), which provided the first high-resolution images of the Martian surface, recording surface details at the two-meter scale, more than an order of magnitude better than previous Viking Orbiter images.

MOC can be thought of as the precursor to the later Mars Reconnaissance Orbiter HiRISE camera that we will discuss shortly. MOC acquired images of Mars with resolutions better than a few meters, thus revealing details about the geological characteristics of Mars unheard of with the twenty-to-one-hundred-meter resolution of Viking. Two meters and below turns out to be at the upper end of where the action is, at least for getting a handle on geologic processes. MOC created a stir in June 2000 when it captured images of what appeared to be small channels or gullies of recent origin in the inner walls of a crater.

The gullies had the appearance of small channels formed by water and one theory held that perhaps these were the result of ground ice in the crater walls or even a leaky aquifer manifesting itself with occasional, short bursts of water at the surface where the crater wall cut across the source layer. But the atmosphere of Mars is too thin to permit permanent flowing water on current Mars, so are these really water-formed gullies? Could they be sand flows, or something else like carbon dioxide gas density flows? While the debate continues about the true origin of these channels, which were later seen in many places on Mars, it has sparked a whole new study of what became known as recurring slope lineae, or RSLs, many of which have formed between imaging of the same locations, which means that whatever the process is, it is still active at least intermittently.

Nanedi Vallis in Xanthe Terra was another interesting image that was widely circulated at the time. This image gives a somewhat oblique perspective of a small part of Nanedi Vallis, which snakes across the old highlands of Xanthe Terra and empties into Chryse Planitia south of the Viking 1 Lander's location. The image reveals a deep valley carved in the Martian terrain. But the really interesting thing was a smaller channel wandering down the floor of the valley. Around this time I was engaged in a small mapping project that was a swath running from the highlands of Xanthe Terra that Nanedi Vallis snaked across and I got very familiar with the valley. Like all large valleys on Mars it emerges practically fully from the highlands without the fine network of valleys feeding into it that we are accustomed to seeing with Earth's river valleys. It is deep and steep-sided, but sinuous, winding its way down to the contact with Chryse Planitia on its north end. And there at its terminus is a faint "river delta" spread out on the low plains. Wow! A delta on Mars.

Mars Global Surveyor also carried a laser altimeter (MOLA) for mapping the surface topography of Mars in detail. The results are a global map of the elevations, often depicted as a bright multicolored and shaded map of the surface. The global MOLA relief map of Mars is pretty much how most of us view Mars these days and certainly gives one a better sense of the landscape than simple optical images, much like a global relief map of Earth helps us to see the varied terrains of Earth a bit better than a global satellite image.

Russian Mars 96. Around 1995, I was standing on a railroad platform in the middle of nowhere somewhere near Potsdam, Germany. A group of us had gotten off the train at what we had thought was the right stop, but nothing appeared right. All we could see were trees surrounding us, and we had no idea where we were, or whether we had arrived at the right place. Fortunately, fellow planetary geologist Steve Squyres from Cornell University figured it out and we caught the next train headed to the intended destination, where we saw the sights. We were on a short jaunt during some downtime while we were at Germany's space research center's Berlin facility, the Deutsches Zentrum für Luft- und Raumfahrt, gratefully more commonly referred to as just DLR, for some early discussions and initial surface imaging targeting for the High-Resolution Stereo Camera (HRSC) to be on board the Russian Mars 96 spacecraft at Mars. This was not long after the fall of the Berlin Wall, so there was construction going on everywhere and traces of the former Soviet era were prevalent. The HRSC was developed by DLR and German planetary scientist Gerhard Neukum was the principal investigator. Neukum was a pillar of the German space efforts, and for many years had been the guru, for all planetary scientists everywhere, of crater counting, the process by which we derive relative ages of the surfaces of planets. He was the classic excitable and extremely serious scholar, a fixture at most international space science meetings, and a member of various spacecraft science teams.

The HRSC instrument was eventually flown on the ESA Mars Express spacecraft and has returned stunning images often displayed in perspective using the stereo terrain models that it acquires as a normal part of the imaging process. But its early incarnation was as the principal Mars 96 camera. Mars 96 launched on November 16, 1996, after a shaky start. The spacecraft itself was enormous and was to be the heaviest interplanetary probe launched up to that time. It was bristling with instruments. In fact some of us sat in the meetings passing around a graphic of the assembled spacecraft with various booms and instrument platforms, including a nice kitchen sink hastily drawn in by some sage. The orbiter had at least twenty-one instruments by my count. And there was also a surface lander, with eight instruments and a penetrator with ten instruments. The spacecraft was so big and heavy that the story was told that during assembly it was affixed to a wall on a support structure for integration activities and that at one point the spacecraft threatened to pull away from the wall and cantilever the whole thing onto the floor.

Launch preparations took place during the early phases of post-Soviet Russia, so things were a little sketchy on some fronts. For example, it was reported, and the details are vague and unsubstantiated, that while the spacecraft was being prepared at the launch site in Kazakhstan, the electrical power to the facility was turned off one night reportedly because the Russian government had not paid its electric bills. As a consequence, engineers were working on the spacecraft at night with gas lanterns. If true, the least that can be said was that they were resourceful and certainly working under nonoptimal conditions.

When Mars 96 spacecraft finally launched on November 16, things were going great right up to the parking orbit, when the fourth stage failed to ignite on its second burn. The spacecraft then separated from the launch stack, and its burn commenced. But failure of the second burn of the fourth stage caused the final spacecraft engine to send the spacecraft back to Earth, where it mostly burned up or dropped into the Pacific. However, parts of the spacecraft were later projected to have landed somewhere in Bolivia, although never confirmed. This last point was somewhat of a concern since the spacecraft was powered by two radioisotope thermoelectric generators (RTGs) based on plutonium-238. At the time we speculated that perhaps somewhere in Bolivia a sheep herder's hut in a remote area was now warmed at night by plutonium RTGs. It makes for an interesting story at least.

Because this was the third Mars spacecraft failure in succession following the Phobos 1 and 2 and Mars Observer failures, Mars scientists were getting spooked. These failures followed a long history of the tribulations of many missions to Mars running afoul of things and prompted a rather famous comment in 1997 by *Time* magazine journalist Donald Neff, that there was a "great galactic ghoul at Mars" that had an appetite for spacecraft. "Galactic ghoul" indeed; it was Coyote Mars, pure and simple. Coyote Mars had gotten miffed at being neglected all those years and was showing us that we could not just start up again with impunity.

The Marsokhod Experiments. Originally the Mars 96 mission was to include a Mars rover, or "Marsokhod," but it was not included in the final mission designs. However, it did provide some early experiments and experiences with construction and operation of a rover for planetary surfaces. The Marsokhod was an interesting fully articulated six-wheeled rover used extensively in the late nineties for field tests here on Earth as a test bed for developing rover controls and interaction with operators in extreme terrains. I was able to participate in an early Kamchatka test remotely during my work at Brown University, as well as later uses at Kīlauea in which we simulated a mission with scientists located remotely at NASA Ames Research Center in California. During the Kamchatka test I was located in the lab at Brown University, where I followed activities via computer screen. The rover acquired images of local scenes with stereo cameras and as images from the rover were returned I pulled up both images and viewed them directly as stereo pairs right on the screen, something that geologists are accustomed to doing, so I suspect that I was one of the few participants actually seeing the terrain in stereo. I was supposed to make geologic commentary on the characteristics of the terrain at the cinder cone where the experiment was taking pace. This being a largely engineering enterprise, the engineers were all amused, as I called out details of the scoria and ash layering as the rover made its way around the site. I was following a protocol for how an astronaut might relay comments on the science in the immediate vicinity of a traverse. It was an early example of doing field geology via a remote rover and that was how I was playing it.

Two missions were simulated during the Kīlauea tests, one for Mars, with the necessary daylong interval between commands and receiving the results of those

The Marsokhod was an interesting fully articulated six-wheeled rover used extensively in the late nineties ... as a test bed for developing rover controls and interaction with operators in extreme terrains.

commands as would be experienced communicating with a rover on Mars, and another mission that simulated a lunar rover with near real-time command and response from the rover. Jayne and I were both involved in a small team working on the lunar simulation in which we were located at NASA Ames and communicated with the rover in Hawaii. The delay via satellite simulated the one- or two-second delay of communicating with a spacecraft on the Moon. The experiment was part of a proposal with the McDonnell Douglas Corporation for a small lunar rover mission based on the Marsokhod rover design to be submitted to the NASA Discovery Program, a series of missions that were low-cost and simple. Jeff Taylor was the science team principal investigator. Jeff was a geochemist who had done a stint at the UNM Institute of Meteoritics back when I was at that university as a geology student. He was an accomplished planetary scientist in the field of meteoritics and lunar samples. But he also had a professional comedian's way of deadpan delivery of quips and retorts that he should have put to use as the official stand-up comedian of science. We had laid out a plan for the mission that would take the rover to one of the volcanically interesting areas of the Moon's near side, and it was going to be a new phase in lunar exploration. But ultimately the proposal was not selected, and we all went our separate ways.

Nonetheless, during the Marsokhod tests we explored many of the problems with remote operation of a rover, getting some early insight into the techniques that would be required. Very early this gave me the idea that one could do terrestrial-style field geologic mapping from a rover. The whole series of experiments was a fun experience paving the way for things to come. More important, the rover bug bit me. That included the excitement of finally doing field geology on the surface of other planets at a human scale of observation. It was an exciting new way to engage in true exploration while making use of all those years of field geologic mapping that I had spent half my career doing with boots and hammers here on Earth. For me it was the perfect blend between my bimodal professional interests in field geologic mapping on Earth and planetary exploration, an odd combination for a planetary geologist.

US Mars Pathfinder. Around 1993 I was attending a small conference held at the Johnson Space Center on the selection of the Mars Pathfinder landing site. This was to be our first attempt to place a small rover, in this case the small kitchen-microwave-size rover Sojourner, on the surface of Mars. The site selection process had not been done since Viking nearly twenty years before this time, yet the tools at our disposal had evolved considerably since then. Since Viking, the technical details of evaluating the surface of a small section of Mars for everything from rock hazards, slopes, amount of dust, and spectral properties of exposures, to of course the likely origin of the particular setting in a geologic sense were finely tuned from years of further developing a knowledge of Mars and remote sensing methods from Viking data. Several meetings took place between 1993 and 1995, in which a dozen candidate sites were presented and discussed, a now familiar process later used in the selection of subsequent landers and rovers.

The selection method starts when a general call goes out to the Mars science community for suggestions of possible landing sites. Everybody likes to participate in a "contest," especially when it revolves around your own specialty and you have something to say or an idea about some exciting place on Mars that you would like to explore up close. But the catch is that the site must meet certain requirements. The requirements for a good landing site generally include:

- certain restrictions in latitude;

- restrictions in site altitude, usually below the mean elevation of Mars, which therefore means somewhere in the northern latitudes;

- restrictions in the roughness and other hazards like big rocks; and

- a geology that meets the mission profile, meaning the need to have local terrain that potentially allows access to materials that would address one of the big questions, like sediments and water in the past.

And, oh, by the way, all this needs to be assessed and the landing must occur within a one hundred kilometer by two hundred kilometer landing ellipse, the zone on the surface where there was some specified high probability that a selected target would arrive, given all sorts of uncertainty in the position of the surface relative to Earth, given the interplanetary targeting capabilities and variations in the atmosphere, which varied the approach as the lander made the final plunge. There are other requirements, but these were the ones that you would need to demonstrate if you wanted to play.

In later years these requirements have multiplied as our knowledge has further grown and our engineering capabilities have been refined. But these were some of the requirements at the time. Nonetheless there was no lack of suggestions. But in the end, we had focused the process to a small area at the mouth of Ares Vallis, which enters Chryse Planitia on its southeast margin, and presented evidence for outwash across the surface. The science reasoning was that since this was to be a mission where we would rove around a small area and look at the rocks with a chemical device, it would be nice if the rocks were a bit diverse. Such a dispersion of rock types was often called a grab bag site. The reason for desiring such a grab bag or random collection of rocks was simply that at a large outwash like Ares Vallis, sourced as it was in the nearby highlands of Mars, we might actually get a batch of different rocks that would give some clue about the diversity of past environments and processes on Mars.

I attended all the Pathfinder site selection meetings, the first one at the Johnson Space Center; a second meeting in Spokane with a field trip to the Channeled Scablands, a possible analog to the outwash characteristics of the top contending site in Ares Vallis; and the final meeting at the Jet Propulsion Laboratory (JPL). In each meeting, there was an attempt to vote the list down to fewer sites, and the winner was the last site standing. In the end all the sites were determined to fit the requirements of safe landing and operation on the surface. But the Ares Vallis site won out on the basis of the most potential for science returns. Much energy, time, and professional resources were devoted by site advocates, so when the hammer came down it was crushing to have your site removed from further contention. But most scientists are no strangers to the process, because it plays out over and over in any career, mainly when you write a proposal for research funding. The proposal process has been likened to doing all the research that you propose to do as a demonstration of what it is you intend to do, yet the proposal may fail to be funded. But that is another story.

Both the landing technology and the rover were part of the smaller and more focused NASA Discovery missions being launched at the time in which the mission, including the launch vehicle, was required to fit under a $200 million cost cap. The Pathfinder mission would be historic, not only because it would be a return, finally, to the surface of Mars, but also because it would be the first, if somewhat modest, rover sent to the surface. Instead of descending the last few kilometers to the surface on a parachute with a rocket braking in the final few meters, this mission would do all of that and then release the lander wrapped in a giant airbag that would bounce, roll across the surface, deflate, and unfold to a perfectly ready-to-go lander. The rover would then roll off the lander and move about the local landscape while relaying its activities back to the lander.

Matt Golombek, a friend and contemporary in the long road to Mars, was the scientist in charge of the mission science team. Matt works out of JPL and has a long history of working with, and often on, JPL-designed Mars landers and rovers. He is an energetic, personally engaging individual who ends most statements with an ironic chuckle. He has made something of a career out of the process of assessing the safety and science of landing sites. He and John Grant from the Center for Earth and Planetary Studies at the Smithsonian have formed a sort of team over the years moderating all the landing site selection meetings. During one trip to JPL prior to the Pathfinder mission launch, I recall Matt taking Jayne and me over to the room where the lander panoramic camera was being tested and calibrated prior to final assembly. The camera was mounted on this mast that popped up upon landing and swiveled about. During the test, I was impressed with how fragile the whole structure appeared. As it went through its panning motion it moved in little jerks that appeared to shake the whole mast like some delicate little toy. But the experience of seeing a spacecraft in front of you is overpowered by the knowledge that this object in front of you will in less than a year be sitting on the surface of Mars. The next time you work with it, it will be a couple of hundred million kilometers away on another planet. The impression never gets old, and every mission since then has elicited the same internal reaction whenever the opportunity presented itself to go see a spacecraft in its birth room.

> [T]he experience of seeing a spacecraft in front of you is overpowered with the knowledge that this object in front of you will in less than a year be sitting on the surface of Mars.

Some of us were invited to the launch of Pathfinder on December 4, 1996, and we traveled to Florida to watch from the VIP viewing area at the launch facility. By this time, Jayne was program manager for the Rhode Island Space Grant (RISG) Consortium, a nationwide NASA educational and outreach program. She was fortunate to work with planetary geologist Pete Schultz, RISG director, faculty member in the Brown University Planetary Geosciences Group, and eminent authority on planetary impact cratering. She and Pete implemented some great ways to use NASA resources to educate teachers and students throughout Rhode Island, including a way to offer the launch-viewing opportunity to two Rhode Island teachers. They accompanied Jayne and me to the launch and it was wonderful to see the experience through their eyes and later see the way they used it to interest their students in science. When the big moment came, the rocket ignited, but instead of a deep rumble of a massive rocket lifting off to Mars I recall it was more of a loud hiss like a small skyrocket on the Fourth of July. But it was a night launch and extremely impressive, and as it arced over on its trajectory the mind was again impressed that this was the start of a journey that would end on another planet.

My time at Brown University was coming to an end, and I had been looking for the next institution from which to pursue Mars exploration and Earth geology, but for those of us from the post-Viking "lost generation" of Mars researchers, pickings were still slim. It seemed that no faculty or research position was available at a university, US Geological Survey, or NASA research center—after all, this was still the time of "no new hires" for planetary geology. So Jayne and I opted to return to at least a Mars-like life in New Mexico, working with the New Mexico Museum of Natural History and Science. Jayne was offered the job of director of education for the museum, but she still owed NASA some Venus geologic maps, and we were both able to continue to do planetary research while also getting back to New Mexico geology. Here we could go on adventures at will amid a landscape every bit as exciting as Mars whenever we wanted to do so. In New Mexico we could experience a landscape that most people could only dream about while studying images of other planets far away.

It was during this time that Jayne and I had invited husband and wife Carl Allen and Jaclyn Allen to Albuquerque to display a Mars meteorite at the museum. Carl was, at the time, curator of astromaterials at Johnson Space Center and Jackie was a planetary geologist and NASA space science educator at JSC. We were very grateful to them for agreeing to bring the meteorite. It received so much publicity and was such a new concept for most folks that we had a couple thousand people move past their display table in the museum atrium to see the amazing rock from Mars. Following the enthusiastic reception of that event, we continued to push for public awareness of planetary exploration by taking advantage of the coming Mars Pathfinder landing and supporting a little more public engagement.

Such an opportunity presented itself during the annual Lunar and Planetary Science Conference in Houston that spring. Jayne and I were both presenting research at the meeting when we ran into another long-term fellow traveler on the road to Mars, Bob Anderson. Bob was on the staff at JPL and was working with the group there on the Sojourner rover and mentioned the intriguing possibility of bringing a working model of the rover to us. So, we invited Bob Anderson and the "Golden Rover" to come to the museum one day for a little public demonstration and he kindly accepted. It was a hit, and our museum visitors got to see how diminutive this rover was and marvel at how it could be controlled from millions of kilometers away on a very rocky Mars. We also worked on our first attempt at a museum exhibit, a small exhibit about Pathfinder that we were able to install at the museum in time for the landing.

Pathfinder went on to arrive in Ares Vallis on July 4, 1997, and operated for eighty-three Mars days, or sols, becoming the first roving mission on another planet beyond the Moon. The landing site as revealed in panoramas from the camera on the lander was a rocky place, but pretty much as predicted by the prelanding analysis, proving that the methods for assessing the surface physical terrain prior to landing using remote sensing methods were right on target.

During its operation, the Sojourner rover clocked over one hundred meters (yes, only one hundred meters, not miles—it was the first time on Mars for a rover) on the odometer visiting nearby rocks and analyzing the chemistry of many of them. The surprise regarding some rocks was the fact that they were a bit more like some Earth continental volcanic rocks in their composition. It is actually a bit more complicated than that, but any discussion about that would quickly become a lesson in volcanic chemistry.

ABOVE: Part of the panorama acquired by the Mars Pathfinder lander. The terrain is exceedingly rocky. The Sojourner rover in this image was a small rover about the size of a microwave oven and the first rover operated on Mars.

At the fall 1997 annual meeting of the Geological Society of America (GSA) after the successful conclusion of the mission, the Planetary Geology Division of GSA invited Matt Golombek to give a plenary talk to the society, and it was a packed room. Matt was from JPL and principal scientist for the project. The Pathfinder mission and its Sojourner rover had been an extremely popular mission with much media hype, so everyone was there to hear a summary of some of the results. I was chair of the Planetary Geology Division at the time, so I had the honor of introducing Matt. I recall that at the end of his presentation I remarked that Pathfinder had put the words "field geology" on the front page of newspapers, to which Matt and the Pathfinder team received a standing ovation. To add to the accolade, we in the Planetary Geology Division had agreed to elect Pathfinder/Sojourner as an honorary member of the Planetary Geology Division.

Nozomi was a Japanese spacecraft that was launched on July 4, 1998, to Mars, a year after the successful landing of Mars Pathfinder. This represented the first attempt by a nation other than the United States or Russia to launch a mission to Mars. Nozomi, which means "hope" in Japanese, was to orbit Mars, studying its atmosphere and

> We were rediscovering that Mars is demanding and only the well-prepared, and sometime redundantly prepared, are welcome.

otherwise developing the technology for future missions to the planets. Among the instruments was a camera that was to image the moons of Mars. Nozomi eventually made it to the vicinity of Mars on December 14, 2003, but control problems plagued the mission during maneuvers to send it into a proper trajectory for orbiting Mars and the mission was terminated. Like the first Russian mission to Mars in 1962, it arrived but was silent. Coyote Mars was still vigilant. The Japanese space efforts have since focused on their extremely successful missions to asteroids.

US Mars Climate Orbiter was part of the Mars Surveyor Program and was implemented as a small focused mission of lower cost. This mission was originally intended to make observations about the atmosphere and also act as a relay for the next spacecraft, a polar lander. Since Mars Climate Orbiter was mostly an atmospheric mission, most of us doing research on Mars's surface had no direct involvement or direct link to the science to be accomplished. Nonetheless the atmosphere and climate have significant influence on the surface and the goals of Mars Climate Orbiter were expected to bring significant dividends to efforts at understanding the evolution of the surface.

It arrived on September 23, 1999, but it too had a problem. I suspect that Coyote Mars, still miffed at being neglected, wanted us to work a little harder if we were to get back into the Mars thing. In this case it was a problem that prevented Climate Orbiter from entering orbit. It arrived safely at Mars and happily communicating with Earth—until it stopped communicating. Later it was determined that it had probably entered the atmosphere and was destroyed instead of slowing down, as it should have. The problem was a software snafu brought about by a mismatch between the engineering units, US standards versus international standards, used for two pieces of software supplied by NASA and by the spacecraft contractor. The glitch meant that the spacecraft thrusters that would have slowed it down and allowed it to enter orbit failed to supply the necessary thrust for braking. Ultimately it was determined that the real failure was the failure to provide cross-checks that would normally find these sorts of issues. It was an aberration in which we lost the touch for a while and Coyote Mars took full advantage of it. NASA got a black eye on that one. But the late-night comedians got some laughs with snarky commentary such as "What did the NASA engineer say when he was caught speeding one hundred miles per hour on a sixty-mile-per-hour road? 'I'm sorry, Officer, I thought my speedometer was reading kilometers per hour.'"

US Mars Polar Lander was designed to touch down in Planum Australe in an area of layered south polar deposits. The goal was to determine whether there was water ice present and to investigate the climate. It launched on December 3, 1999, and just a little over two months after the Mars Climate Orbiter disaster, it too had a problem. As the spacecraft was completing the descent to the surface on December 3, 1999, communications were lost. Later review determined that the likely cause was a false signal during deployment of the landing legs as it approached the surface effectively telling the spacecraft that it had landed, shutting the braking rockets down, and resulting in a high-speed impact on the surface. It was a failure of the design process since the potential for a false signal was known but not accommodated in the final hardware execution. After several recent mission failures, it was beginning to appear as though we had lost our "touch" in designing successful Mars spacecraft. But in reality many fault the exploration architecture of the time that relied on the motto "faster, better, cheaper" in which mission cost would be reduced by focusing the missions on simple goals and using available spacecraft designs and components rather than expensive one-off designs. We were rediscovering that Mars is demanding and only the well-prepared, and sometimes redundantly prepared, are welcome.

The desire for information about the poles of Mars was not forgotten, however. Eventually many of the instruments or later versions of those investigations, including a panoramic camera somewhat similar to the one on the Mars Pathfinder lander, were later flown on the successful Phoenix polar lander.

US Mars Odyssey, which launched on April 7, 2001, and arrived in orbit on October 24, 2001, was thankfully successful and has become a workhorse of Mars exploration. When it went into orbit it quietly set up shop and began changing the perspective we had of Mars. Few would have guessed at the long life that lay ahead of it, nor the sheer volume of results. A simple summary of Mars Odyssey would be an encyclopedic task because of its many accomplishments and because it is still continuing; as I write this, it is currently the oldest surviving spacecraft in Mars orbit, having continued to function and return important information for two decades. In addition, it has served as a relay satellite for four surface landers and rovers. We used Odyssey regularly on the Mars Exploration Rover mission during its daily pass in the skies of Mars. It was our go-to for relaying data to Earth.

Yes, the mission name was a play on the movie title *2001: A Space Odyssey*, based on a science fiction novel by Arthur C. Clarke. But it had a previous preliminary name, ARES, acronym for Astrobiological Reconnaissance and Elemental Surveyor.

But the name was too ambitious-sounding for many people and the name Odyssey was resurrected from a list of suggested names in the end after Clarke said he felt it was a great name.

On an obliquely related side note and while on the topic of names of missions, it is amusing to point out the recurrence of some names used for many missions over a spread of several years throughout a broad spectrum of society and business. For example, the word "pathfinder" appeared to be all the rage in the late nineties. There was an SUV named Pathfinder, I recall an off-road tire named Pathfinder, and there were several other uses that are lost to memory. But it was a word that appeared frequently. Then it dropped out and the word "odyssey" appeared to be the new favorite. There were various products named Odyssey, from electronics to minivans. Spirit appeared for a while in the first decade of the 2000s. There was even an airline by that name. Lately "endeavour" is the hot ticket, following the appearance of the crater of that name from the end of the Opportunity mission. "Curiosity" is a recent favorite, with several products, media outlets, and probably other things bearing that name following the Curiosity rover usage of that name. I wonder if this trend will persevere?

One of the important instruments on Odyssey was headed up by Phil Christensen at Arizona State University, a camera designed to map the spectral characteristics of the surface using thermal imaging spectrometer (Thermal Emission Imaging System, or THEMIS). Phil had been hammering on the use of near-infrared imaging for years, and he would later become the principal investigator for the Mars Exploration Rover Mini-TES instrument. But the Odyssey TES would look at the entire planet at moderate resolutions. The importance of this was the fact that near-infrared creates clear, crisp, and less hazy images, so for those interested in the physical details of the surface the effects of the atmosphere would be less. Also, it would permit a global look at the distribution of rocks versus dust, which would benefit future landing site evaluation among many other things. THEMIS is responsible for many of the modern regional images of Mars used frequently in this book.

An exciting new instrument at Mars, also on Odyssey, was the Gamma Ray Spectrometer (GRS) designed to determine the elemental composition of the surface and detect the presence of water and water ice. It did this and more. One of the first widely circulated results was the detection and global mapping of water. The GRS instrument detected elements by making use of natural cosmic rays impacting the surface and scattered a variety of gamma rays, of various energies back. GRS simply looked for those gamma rays and using knowledge of the interaction of high-energy cosmic ray particles with matter, it would assess the abundance of different elements. The interaction occurs in the upper meter or two of the surface, so the nice thing about GRS is its ability to "see through" any surface covering like dust and detect the presence of things in the upper part of the surface, including, for example, water ice just below the surface or water molecules that are bound to the minerals.

ESA Mars Express/Beagle 2 Lander was the first Mars mission attempted by the European Space Agency (ESA). The Mars Advanced Radar for Subsurface and Ionosphere Sounding orbiter Mars Express has been a resounding success and is now the second longest mission operating at Mars after the US Mars Odyssey spacecraft. On board are a variety of instruments, including Visible and Infrared Mineralogical Mapping Spectrometer (Observatoire pour la Minéralogie, l'Eau, les Glaces et l'Activité; OMEGA), Mars Advanced Radar for Subsurface and Ionosphere Sounding (MARSIS), and High-Resolution Stereo Camera (HRSC), and five other instruments designed to understand the space and upper atmosphere environment at Mars. It hit the brakes and successfully arrived in orbit at Mars on December 25, 2003. Shortly before arrival it also released the British lander Beagle 2, targeted for the surface within Isidis Planitia. At the time, I confess that many of us, especially those of us just about to head to JPL as science team members for the NASA Mars Exploration Rovers Spirit and Opportunity rovers, were not following the event closely because we had other issues on our minds.

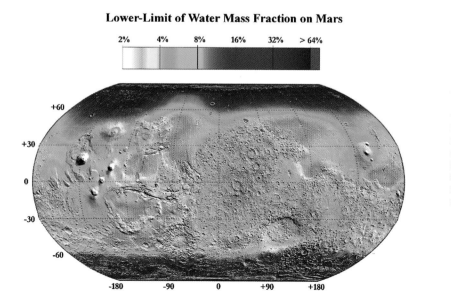

Lower-Limit of Water Mass Fraction on Mars

2% 4% 8% 16% 32% > 64%

LEFT: Global map of the estimated lower limit of water content in the upper meter of the surface of Mars from the Gamma Ray Spectrometer, Mars Odyssey.

Mars Express has been a workhorse for the ESA. But the OMEGA, HRSC, and MARSIS instruments have generated most of the publicity for the mission. OMEGA's ability to detect various minerals of importance in resolving the geologic history of Mars has been important in the quest for evidence of past water on Mars. The fact that OMEGA was mapping most of the phyllosilicates, a clay mineral, in the cratered and ancient terrains of Mars prompted Jean-Pierre Bibring and colleagues on the OMEGA instrument team and related instruments teams on other missions to propose an alternative geologic time scale for Mars, as was discussed in the previous chapter.

Because the HRSC is a stereo camera, the high-resolution and colorful images acquired by Mars Express are often presented in three-dimensional relief, resulting in some spectacular perspectives on the geologic landscape of Mars. MARSIS, the subsurface radar sounder, generated a stir when in 2018 it was reported to have detected a substantial body of water 1.5 kilometers beneath the south polar cap. Subsequent detections identified additional lakes in the subsurface of the south polar cap.

BELOW RIGHT: The Beagle 2 lander as it would be deployed on the surface of Mars.

BELOW LEFT: MRO/HiRISE image of the Beagle 2 lander on the surface of Isidis Planitia, Mars. The lander was only a meter across. Close inspection of the image shows that the solar panels were only partially deployed.

The Beagle 2 lander was the brainchild of British planetary geologist Colin Pillinger at Open University and several colleagues. Its goal was to analyze Martian surface and subsurface materials for chemical signs of past life. The name Beagle 2 of course was in honor of the ship that had taken Darwin on his great biological mission of discovery in the 1830s. It was to land in Isidis Planitia on

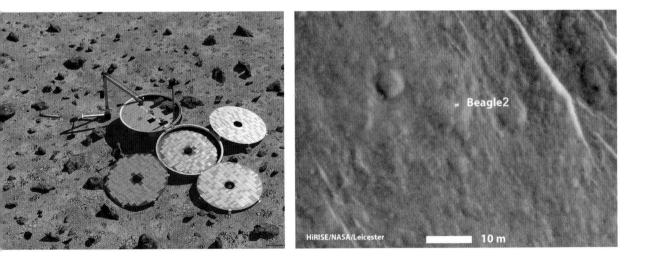

December 25, 2003, just a week before the first Mars Exploration Rover Spirit. But after separation from Mars Express and the commencement of the entry, descent, and landing sequence, it was never heard from again. At the time it was presumed to have crashed, and several inquiries suggested various technical glitches that may have caused the landing to fail.

However, in 2015, Mars Reconnaissance Orbiter high-resolution camera HiRISE finally spotted Beagle 2, eleven years after its loss. The resulting images showed that Beagle 2 had in fact successfully landed. There were no scattered parts and debris such as that previously seen from orbit around other failed landers. Instead Beagle 2 was sitting on the surface intact and upright as planned. The problem appeared to be that two of the four solar panels failed to deploy and were covering the communications antennas. Sadly, Beagle 2 survived and may have died only after its batteries ran down. Thus the engineering behind the lander descent and landing system was shown to be successful and not at fault in the mission failure. The heartbreak about this discovery in 2015 was the fact that Colin had died eight months previously. He would have been delighted to know that his mission had actually made it to the surface in one piece but was thwarted by a technical glitch, at the last second, as it was about to begin its activities on the surface of Mars.

US Mars Exploration Rovers (MER) Spirit and Opportunity launched on June 7 and July 10, 2003, respectively and was an ambitious program to finally put two capable rovers on the surface of Mars. I will discuss them a little here to place them in the context of the timeline of missions but go into much greater detail later in the book. I had eagerly awaited these rovers for many years and it seemed to me that most of my professional life had been a lead-up to some sort of mission in which a rover would actually traverse the surface and see the geology up close just like we geologists do in the field on Earth. So I naturally wrote up a proposal to be a member of the Participating Scientist Team for this mission. In late 2002, after the proposal review process had been completed, I got the "call." I was on the mission, and what a mission it was. The first real "in person" exploration of the surface of Mars.

US Mars Reconnaissance Orbiter (MRO). It was a cool day in northern Arizona, and I had hiked about halfway up the steep western slope of SP Crater, a beautiful young cinder cone in the northern reaches of the San Francisco volcanic field near Flagstaff, Arizona. I had gotten a head start because I knew that I would be overtaken quickly by some of the other HiRISE science team members as we ascended the cone

during a little field trip. Not because I was slow, but because, as a field geologist surrounded by remote sensing geologists, I was too busy looking at the rocks at my feet.

Although it was still in the early days of the MER mission, I had also successfully applied for a position as participating scientist on the MRO/HiRISE team a year or two before, so I was there in the capacity of a science team member. The science team like all mission teams was having its annual science team meeting and during each meeting, the HiRISE team was accustomed to having a field trip ostensibly to see real geology to help support the work we were all doing with Mars geology using the high-resolution images being acquired by the HiRISE instrument. Sure enough, shortly after I had stopped for a breather, here comes Al McEwen, the principal investigator for HiRISE, and a host of students storming up the slope. They blasted by me and continued without stopping to the rim far above. Now, as a field geologist, I have climbed my share of volcanoes and I am not particularly out of shape, so it is important to note that if Al were the metric, everyone would be considered out of shape. An avid jogger, he has the lanky shape of most committed runners. And clearly a quick climb of a 30-degree slope populated with ball-bearing-size cinders was just a nice little workout for him. I believe that he considered these trips to be a group-bonding-exercise event, more than a geologic field trip.

The funding for the Participating Scientists program dried up a year later, as funds for missions tend to get squeezed as time progresses for each mission, but not before I got to be a part of the MRO family. MRO was one of those missions whose results go on and on, and the results and data products become so much a part of the current science that it is difficult to remember a time before there was an MRO.

MRO launched on August 12, 2005, and arrived at Mars on March 10, 2006, loaded with an array of cameras and remote sensing instruments that have written the new geography of the red planet. HiRISE was the largest camera ever sent to Mars and acquires images with pixel resolutions of twenty-four centimeters, which is better than the approximately one-meter resolution that we are all accustomed to with typical Google Earth data.

Each image is a world in itself with so many things to explore. Like looking down on your city from above or opening a dictionary and getting lost in one page, you can be lost in one image. Whole books could be done just showing the individual images. In fact, whole books have been done. But usually only a small part of one image can be adequately displayed in anything approaching the resolution and detail captured in these enormous images.

In addition to the HiRISE camera there is the Context Camera, or CTX, with a wider field of view designed to provide the larger view for the areas targeted with instruments like HiRISE. With a resolution of six meters, CTX images exceed the best Viking orbiter images and are now a staple for anyone attempting to present a regional picture of a particular place on Mars. Like Odyssey's THEMIS camera many of the separate images have been mosaicked to provide even larger area views.

There is also a global camera, MARCI (Mars Color Imager) somewhat like the full-disk Earth cameras we use in weather satellites. In fact, dozens of images are acquired on a daily basis and form a modern global weather satellite type of view for Mars. It is with this camera that important "weather events" on Mars, like regional

BELOW LEFT: Some lava flows in the Elysium region are so young that there are almost no impact craters, which means that they may be less than a million years old. This example is an area one kilometer across. **BELOW CENTER:** HiRISE image of the complex layering in the floor of Melas Chasma. **BELOW RIGHT:** MRO/CTX images cover larger areas than those acquired by MRO/HiRISE. CTX has revealed some interesting medium-scale volcanoes. An example is this small shield volcano with a small summit crater on the lower east slope of the Tharsis region between Pavonis Mons and Noctis Labyrinthus. Note the later lava flows that have flowed onto the lower slopes in the upper left.

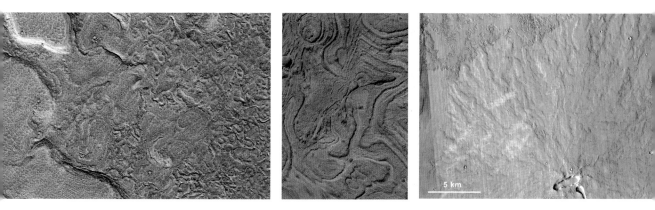

global dust storms, are identified and tracked. These provided an early warning system on many occasions for the rovers operating on the surface.

CRISM (Compact Reconnaissance Imaging Spectrometer for Mars), another of the instruments on MRO, has been another staple of the Mars Exploration Program. CRISM's unique capability is the detection of minerals at fairly high resolution as spectrometers go, about twenty-five meters. It has been the go-to instrument for determining the presence of minerals important to the quest for evidence of past aqueous environments, such as clay minerals and carbonates. But CRISM has also detected things like hydrated sulfates. It has been used to assess all landing sites proposed since it became operational in orbit and it has further aided in mapping and understanding many other regions of interest in the overall exploration process. We also used CRISM like a divining rod on the Opportunity mission for finding sites with interesting minerals to be visited by the rover.

SHARAD (Shallow Subsurface Radar) is able to use radar to probe beneath the surface in search of ice, particularly in the higher latitudes where it is predicted that water ice should be stable fairly close to the surface. But it has also been able to map the thickness of the polar caps. Results of orbital traverse across the north polar ice cap show that it is about two kilometers thick, including four distinct layers less icy that may represent the cycles of climate change associated with variations in the tilt and orbit of Mars.

US Phoenix Polar Lander. On May 25, 2008, the day that the Phoenix polar lander successfully set down on Mars, at a latitude of 76.7°N, I was upstairs in the building devoted to HiRISE science planning on the University of Arizona campus. I had been quietly immersed in targeting-related activities for the HiRISE camera.

Because HiRISE was headed up by Al McEwen of the Department of Planetary Sciences at University of Arizona, the operations were supported in a new building across the street from the department's Kuiper Building where I had worked as a PhD student. I was there because those of us on the HiRISE science team rotated through the position of targeting specialist and worked on site at the University of Arizona mission control during that time. My mission duties happened to coincide with the day that Phoenix landed, so I was not terribly surprised that the whole campus was abuzz with the event. The Phoenix science team had their own building off campus for spacecraft instrument operations. But on this day a sizable group had gathered on campus at the HiRISE facility. I was sitting upstairs in a mostly empty science area working on some of the latest targeting for the HiRISE camera, and as the room got quieter and the time of the Phoenix lander drew near I began hearing a faint hubbub downstairs, so I decided that I had better walk down the stairs and see what was happening.

The lower floor was occupied by a large atrium devoted to displays of a mock-up of the HiRISE camera and several monitors with the most recent images. There were several monitors also attached to the ceiling. It was on these monitors that the spacecraft operations room at JPL was being displayed in anticipation of the landing. As I stepped off of the stairs, I was overwhelmed by the loud crowd that had gathered in the atrium; the crowd spilled out the front door and onto the sidewalks and central mall of the campus. It was no wonder that the HiRISE science operations rooms on the second floor were empty. Everyone was downstairs. The lander was in its final approach, and, as the entry, descent, and finally the landing happened, the crowd

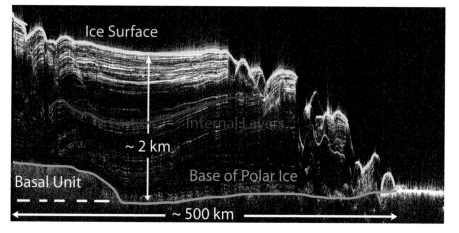

LEFT: MRO/SHARAD radar-sounding profile results across the north polar cap. The cap appears to be about two kilometers thick. The profile is 250 kilometers long, and the vertical scale is greatly stretched for visibility.

exploded. Another successful Mars landing had taken place, and it was in a place that we had not previously experienced on Mars, a high-latitude site near the north pole. As the atrium emptied out, I went back upstairs and continued slaving on the HiRISE target selections.

Just before the Phoenix landing we had discussed the possibility that HiRISE might attempt to image the lander as it was descending to Mars. It sounded like a crazy idea, requiring that the orbital camera be at the right place to get a view and be there at precisely the same time as the landing. But as it happened MRO with HiRISE aboard would actually be overhead near the Phoenix landing site when the landing took place. It was a long shot, but an imaging sequence was planned just in case. We did not expect a good view as the Phoenix spacecraft and parachute would only be a few pixels wide in the enormous field of the HiRISE camera, but it was worth a shot. The next day after the landing, when the HiRISE image that would include the landing scene came down, many of us pored over the image to see if there were any indications of the lander. Since most of the HiRISE images are twenty thousand pixels by four thousand pixels, there were a lot of pixels to scan through during the search.

We were stunned. In one corner of the image that was looking obliquely across the landscape with a large deep crater, Heimdal Crater, in the background, was a bright white fully inflated parachute with a bright white speck of the Phoenix lander dangling beneath. We had captured the first-ever image of a spacecraft in the act of descending toward the surface of Mars.

The planetary science group at the University of Arizona has consistently fielded some innovative instruments on planetary missions and was also becoming a designer of entire spacecraft missions. The Phoenix polar lander was one of them, and the principal investigator for that mission was Peter H. Smith from the University of Arizona optical sciences department together with twenty-four other co-investigators from institutions around the world.

The goal of Phoenix was to investigate whether water ice was present at the extremely high latitudes of Mars. Predictions from what we knew at that point were that ice should be present very close to the surface, perhaps just below the surface cover. And so it was, as water ice was seen on the surface beneath the lander where the landing rockets had blown off the soil. Later the lander would dig a trench exposing ice in the subsurface and thus became the first mission to actually see and touch water ice on Mars.

Phoenix went on to make many other unique observations regarding the chemistry of the surface, in particular the presence of certain salts that would not exist

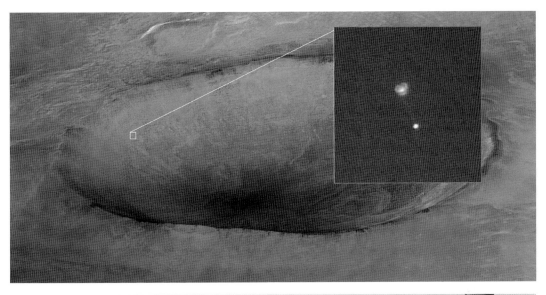

TOP: MRO/HiRISE image of the Phoenix lander on its parachute descending to the surface of Mars. The ten-kilometer crater Heimdal is in the background about twenty kilometers away. Inset shows a detail of the descending spacecraft. **ABOVE:** The Phoenix landing site acquired by the Surface Stereo Imager looking west during Phoenix's sol 16.

ABOVE: View underneath the Phoenix lander by the robotic arm camera on sol 142 showing a patch of ground ice exposed by the landing rocket blasts.

had any form of water been present in the past several hundred million years. So the conclusion is that the environment is extremely dry despite the presence of water ice in the ground. The ice apparently has not attained the liquid state in recent Mars geologic history. When the ice at the Phoenix site was exposed in the trench it gradually sublimated away.

Phoenix was solar-powered and designed to last about three months on Mars before the onset of northern hemisphere fall. It continued for another two months beyond that but eventually went into a low-power state as the Sun dipped ever closer to the horizon. Although the possibility was considered that Phoenix might wake up and communicate as the sunlight returned in the next spring, the near polar conditions were against its survival. During the Martian northern hemisphere winter, the Phoenix site is within the extent of the seasonal carbon dioxide (dry ice) polar cap. The thickness of dry ice snow predicted is small by Earth standards, around fifteen centimeters; dry ice is dense and the buildup on the solar panels would likely be too great for the structures supporting them. Later HiRISE images in the following year showed this to be the case.

MARS EXPLORATION PHASE 2 ENIGMAS: QUESTIONS AND MORE QUESTIONS

MANY SCIENTIFIC PAPERS AND ENTIRE BOOKS have been written outlining what we have learned about Mars from the missions discussed above. The study of Mars has exploded exponentially in the past two decades since the return to Mars with more capable spacecraft and instruments. Nothing short of a Mars encyclopedia could truly adequately attempt to represent the rich history, the results of each mission, and the new questions arising from each. Nonetheless, a few areas stand out and are the subject for scientific debate or were widely discussed in science and popular media outlets. Many of these topics are developing into entire fields of endeavor in their own right.

One topic that has grown over the past two decades is evidence for water on Mars in recent times. The appearance of strange streaks on the walls of craters first observed by the Mars Global Surveyor MOC has continued to be a topic of study, now further expanded with high-resolution images from the MRO/HiRISE instrument. The dark streaks appeared to be relatively recent disturbances that had the appearance of sudden bursts of some fluid at the upper slopes of deep craters that streaked down the interiors. One interpretation was that the streaks originated when a layer of ground ice in a frozen near-surface aquifer was released and flowed down slope. Others felt that simple dry landslides could do the same thing. The whole topic became a source of some concentrated interest. Continuing the need to speak of things with names that do not presuppose an origin, these streaks became known as recurring slope lineae, or RSLs. The Phoenix polar lander had shown that water ice was present in some areas, as has the detection by MRO/CRISM of water-ice rich layers in the scarps on the margins of the polar caps, so the idea of a cloud of water vapor, or perhaps salty water more stable at the low atmospheric pressure and cold temperature of current Mars, released from exposures in crater walls was not off the table. The topic remains an area of study.

Observations by ESA's Mars Express and more recent observations by the Curiosity rover have shown that methane is present in the atmosphere. Given that methane does not hang around very long this implied some sort of active modern-day release was involved. But is the methane of a strictly inorganic origin, which is entirely possible, or is there some microbial origin afoot? The detection of methane itself remains sporadic and the abundance and recurrence itself is just beyond the threshold of understanding. It remains a big question that current missions are as yet grappling with.

The possibility of modern volcanic activity on Mars remains a question. Study of the surface using HiRISE images of lava flows in the Elysium volcanic area, particularly in the Cerberus Fossae area, has renewed ideas that not all Martian volcanism was something that happened in the deep past. Many of the lava flows in the Cerberus Fossae vicinity have so few craters and have such fresh-appearing surfaces that ages of just a few million years are estimated. In addition, there is evidence that Cerberus Fossae itself was the source of a relatively recent large outflow of water that carved the surface contemporaneously with volcanism. Did the magma erupting from the fissure of Cerberus Fossae encounter and melt shallow ground ice? This is another area of current scrutiny that has the potential of changing the view of Mars as a place where internal geologic processes are no longer significant.

Several of the recent orbital missions have identified carbonates and clays among the older terrains of Mars. Both minerals require a water environment in which the water is more neutral than the very acidic conditions that were later associated with sulfates. So this has encouraged thinking that the early geologic history of Mars was a time of environments more conducive to "habitability" by microorganisms if Mars ever had life at all. This has supported a Mars exploration theme that seeks evidence of possible past life. Both Curiosity and its successor Perseverance have continued the quest for organic materials on Mars. Meanwhile entire programs of study in the rapidly developing and expanding field of astrobiology are exploring the ways in which life can now exist or could have in the past existed on Mars.

There are two directions of study in the quest for possible life on Mars. One seeks to look for evidence of past life in the geologic record. That is the one now familiar to us all from the story of life through time common to many natural history museums. There is plenty of evidence of sediments from the time of abundant water on Mars, and the search would include fossils and their microbial equivalent. The other direction of study asks how and where life might exist today. This question is usually couched as an assessment of ways that life can live underground or in rocks and of ways that existing environments might retain materials necessary for sustaining life. An offshoot of this direction of thinking includes the ability of life to flourish in caves. In the case of Mars, caves would offer protection from radiation that drenches the Martian surface. We now know that the open pits in the volcanic areas are direct evidence for the existence of caves. The environments presented by such caves are potential places where any extant organism could flourish.

The stage was set by this point for getting down and putting boots, metaphorically speaking, or wheels, actually, on the ground. And so the next phase of Mars exploration began.

Many other questions had been formulated about the abundance of water and the climatic variability of Mars as a result of our expanded knowledge about Mars in the past few decades, most of which were waiting to be answered. This was the state of our knowledge, and our questions, prior to the next major mission, the Mars Exploration Rover. This would be a return to the actual surface. But, unlike the Viking Landers or Pathfinder, where our view of the surface had started, this time we would be given the opportunity to move around and see more than the limited view from a stationary perspective. Many of the questions that came from the initial views from above with orbiting spacecraft could and had been addressed with new orbital missions with increased resolution and better spectral instruments. But we simply needed to see the ground up close and in more than just the couple of places offered by the Viking Landers to begin answering those questions.

Mars exploration up to this point had peeled back layers of mystery about the planet, crushing old mysteries but revealing new mysteries that begged to be addressed as all exploration is wont to do. The view of Mars had gone from a lively world peopled by a civilization capable of engineering giant canals quickly to a dead dusty, crater-pocked planet, and then just as quickly back to one that had clearly supported, if not civilizations, at least an environment with ancient rivers and landscapes carved by erosion. From the very beginning of modern spacecraft missions, primarily the Mariner 9 mission, it was clear that water was an important part of Mars's geologic history. And so, the question of life on Mars was resurrected and became the theme of the first landers, the Viking Landers of 1976. Nothing that we have learned about Mars since then has altered that theme. In fact, everything that we have learned since then has spurred us on to look more closely or with better investigations. The search for evidence of life—past or present, fossil or extant—has been built around the theme of water; its past occurrence as liquid, ice, or vapor; its abundance and longevity on the surface or beneath it in the ground; and its effects on everything from landforms to minerals.

Because the history of water was a central theme to many of our questions about Mars, by the late nineties NASA's programmatic plan for Mars exploration was crystalized as a pursuit of water's role in the geology and potential biological development of the planet. So the theme "follow the water" had emerged as the one connector in all of the programmatic goals.

Up until this point we had benefitted from the fact that the geology of Mars is a particular type of geology that was exposed and favorable to being looked down upon from orbit. From this overhead view you could see the landforms such as individual lava flows or canyon walls clearly layered with stacks of geologic units. The shape of the landscape itself testified to the rock types in a general way, at least in a way that allowed some reasonable guesses about the geologic materials making up the surface and the geologic history of the surface. It was clear that the next step was to get on the surface and do some human-scale exploration if we were to understand the geologic history of Mars better. The stage was set by this point for getting down and putting boots, metaphorically speaking, or wheels, actually, on the ground. And so the next phase of Mars exploration began.

PART

2

ROVING A NEW WORLD

SEVEN

Wheels on the Ground

"An optimist will tell you the glass is half-full;
the pessimist, half-empty; and the engineer will tell you
the glass is twice the size it needs to be."
—OSCAR WILDE

SOMETIME IN THE MIDNINETIES, I WAS sitting in the conference room at the hotel across from the Johnson Space Center in Houston, and I was about to hear the next of umpteenth presentations in a small conference about future Mars rovers and Mars sample-return mission concepts. Rovers on Mars had always been a dream mission for the Mars exploration process. Even as the Viking Landers were just beginning their missions on the surface back in 1976, one idea that was highlighted in a mission newsletter suggested that the next step would be to put tank tracks on each of the footpads of a Viking Lander so that it could move around instead of just sit in one place. Yet many years were to pass before the idea of rovers on Mars began to seriously emerge in discussions. Granted that no rover had been to Mars at that point and our view of Mars was still pretty much what we got from Viking, that is, it was a kind of one-hundred-meter resolution view of Mars. We knew the big features, but we knew too little about the small landscape stuff that you and I would notice if we were to be plopped down on Mars. But in the midst of this vague understanding of the scope of interacting with a terrain about which we knew little, we still needed to discuss the needs of future missions.

In the particular conference that I was attending, the next presentation was from an engineer responsible for early thinking about how we would control rovers on Mars, including autonomous behavior and hardware requirements of the rovers that would navigate the terrain and collect samples for return to Earth. The issue, of course, was the simple fact that Mars is 250 plus or minus 100 million kilometers distant from Earth, so a radio signal—for example, a radio signal controlling said rover—would take up to twenty minutes . . . one way. The delay between saying "stop" and the time it might take to arrive at a cliff was hardly conducive to using a joystick. So the early thinking was evolving to the concept that we might want to invest rovers with certain autonomous behaviors and restrict out commands to simple statements, like "Go to point B, stop, and take a panorama," that sort of thing. The rover would figure out how to get to "point B" and if it couldn't, it would not do it and wait for more detailed instructions if necessary.

During the presentation by the engineer, one of the other scientists in the conference room noted that we will want to do some field geology with the rover too, examining outcrops, collecting samples, and basically doing the things we think are important when trying to understand geology here on Earth. So someone just asked, "Can we do geology with a rover?" The engineer did not hesitate one second and responded, "Yes! You tell us how you *do geology*, and we will program the rover to do that."

The room, filled mostly with planetary geologists, was dead silent at first, followed by a few snickers. Of course, the geologist version of what he just said was like telling a surgeon, "Yes, we can have the rover do surgery. Just tell us how you do surgery." "Doing geology," or surgery, is not a set formula. It is research or a finely honed skill after all, each project (surgery) is different as is each outcrop (individual), and each project develops from the application of that expertise at every turn. Geology is something that happens after application of all sorts of expertise, not something that you automatically flip "on" or "off." Geology is something that emerges after you do the things that geologists need to do to determine the geology of some place. Aside from all that, I don't think any modern geologist ever considered the word "geology" to be a verb, although I have noticed a tendency in mission-speak to refer to the process of "doing science" or even "geologizing," which presumably means doing science activities instead of spacecraft-related systems activities.

It was not until many years later, during the first days of the Mars Exploration Rover mission, that it became obvious that the engineer really did know what he was talking about and we were the ones with the wrong idea. He was not asking how you do the science of geology, he was not asking us for the formula for investigating something, he was looking at it from the system-engineering perspective at the "granular level" of actions from minute to minute, that is, the process of doing the science of geology in a given task. He was asking, "What are the activities that you do and observations that you make when you do geologic fieldwork?" to get the data that you need in order to arrive at geologic conclusions about a place.

BIRTH OF AN EXPEDITION

THE MARS EXPLORATION ROVERS CAME INTO being during the consolidation of the emerging views of Mars that had occurred during "phase 2" of our invasion of Mars over the previous two decades. Particularly after events of the late nineties, namely the Mars Climate Orbiter and Mars polar lander fiascos, NASA was considering the next steps in efforts to recover the science goals of those missions. The thinking was continuing to evolve about what those steps might be. One was a new

> "They say a name expresses the thing it stands for, but I wonder if it
> isn't the other way around—the thing gets more and more like its name."
>
> —HARUKI MURAKAMI, *The Wind-Up Bird Chronicle*

orbiter mission with advanced instruments to do most of the science related to understanding the climate history of Mars, particularly this question about the history of water, and more to the point, the history of potential habitability by organisms in the past. But many of us felt that all this remote sensing, looking down from above, was only going to take us so far in getting at the answers. There was a growing desire to just get down to the surface and look at the rocks up close.

Like all missions, it is surprising that it happened at all and only then after a prolonged and somewhat agonizing gestation that wended its way through the harrowing process of mission proposal and ultimate execution, with sudden death looming at each phase of the process.

The Mars Exploration Rover (MER) project was headed up by two impressive scientists, and it was largely because of them that the mission was so successful and fun. Steve Squyres of Cornell University, principal investigator for the Mars Exploration Rover instrument package and science team, was one of those people said to have brains that operated on faster processor speed than us normal humans, favored cowboy boots and jeans, and was thought by many to have the presence of a rock superstar. Ray Arvidson of Washington University in St, Louis, deputy principal investigator, was also a laser-focused planetary scientist with a dry delivery that crackled with an underlying intense understanding, analytical skill, and knowledge about whatever science topic he talked about. His dry wit would pop into his presentations at unexpected moments. Both had a vision of putting a bunch of instruments on a mission to the surface that would simply replicate the principal tools of a field geologist, one being the mobility that can enable moving over the surface and examination of different rock outcrops. And the farther the movement, the more rocks that can be examined. Perhaps more important, as I had found from field geologic mapping in relatively uneroded terrains, the farther you move across a landscape with a few nice natural cross sections from erosion, the more geologic units or different types of geologic formations you can encounter and examine. Hence the plan was a suite of "geology field" instruments on a rover capable of moving far beyond the local setting of the landing site, something that had not been done before.

There was another important reason why the rover mission was different and many of us believe that it was an important reason not only for its success, but why it was so enjoyable. The culture of Mars surface-mission science has for a long time been all about instruments that make specific observations, looking at soils, analyzing the chemistry, and measuring properties in spectra. This has meant that mission teams for most planetary missions have consisted largely of a collection of instrument teams, each instrument team being responsible for one instrument's care and feeding, that is, running it and making certain that it gets used to maximum benefit for the mission goals. As a result, most planetary missions consist of a confederation of instruments, each with their individual principal investigator and science teams. The mission is then operated such that the instrument principal investigators are the ones who meet at the roundtable of daily operations. The followers in each kingdom tend not to interact at the mission level unless appointed by the principal investigator to do so. As a rule this means that the mission is an exercise in making good use of the instruments.

The MER mission took a different approach. There were no separate instrument teams. Instead the rovers were themselves the instrument and the science team was the instrument team. Steve organized this whole package of instruments in a single group named the Athena instrument team. This simply meant that every member of the science team was functionally equal to any other member and could come to the daily operations roundtable with input. Steve's management style, and that of his deputy principal investigator Ray Arvidson, fostered a happy and involved team that worked well together.

The mission as originally conceived by Steve consisted of one rover, but it became two rovers by a happy accident. It is a long story, but it came down to a NASA headquarters decision to hedge the bet that we could get a rover to the surface safely by rolling the dice twice. That is the short version. It also just so happened that the "production line" of assembling two spacecraft proceeded in series, and the first rover was given the assembly name MER1 and the second to be assembled was MER2. Due to the vagaries of assembling and launching twin spacecraft, MER2 was completed and sent to the launch facilities first and became known by the mission project name MERA. By the same chance, then, MER1 became MERB. The first to launch would be the rover going to Gusev

Crater and that was on the books as being the rover given the name Spirit. So MER 2 became MERA, and MERA became Spirit. Henceforth the names Spirit and Opportunity were off-and-on referred to through the mission by the codes MERA and MERB almost interchangeably, and so the project names MERA and MERB appear in various archival documents of mission activities that followed.

The names MERA and MERB were of course the project designations, but, as I said, the rovers were given names. Naming followed the tradition established with Pathfinder's Sojourner rover. A call went out for students to enter a competition for naming the rovers and a committee culled through the ten thousand proposed names and accompanying essay explaining the proposed names to determine the most appropriate. The winner was Sofi Collis, a third grader from Scottsdale, Arizona. Sofi was born in Siberia but came to America when she was adopted by American parents. In her essay she explained the proposed names: "I used to live in an orphanage. It was dark and cold and lonely. At night, I looked up at the sparkly sky and felt better. I dreamed I could fly there. In America, I can make all my dreams come true. Thank you for the 'Spirit' and the 'Opportunity.'"

I think most of the team had mixed reactions to the selected names. A general sentiment was that those were attributes, not real names. But the names grew on us and eventually we would not have it any other way. Thus began a tradition used since then of naming rovers, and now Martian helicopters, with attributes, such as Curiosity, Perseverance, and Ingenuity.

As the mission of the two rovers progressed, we learned that names could have some influence, or so it seemed. Opportunity had all the opportunities and Spirit showed great spirit while facing significant adversities. Opportunity just did the job, while Spirit occasionally had some great drama that we all had to suffer through. The lesson of the story might perhaps be to be careful what you name a mission, particularly if you use an attribute instead of a common name, because it appears that the attribute can become a characteristic of the mission. With the following missions, Curiosity sounded safe enough and has been so, but Perseverance sounded like it could get a little edgy, and it was with preparations to launch during a global pandemic.

The idea of designing the Mars Exploration Rovers to be robotic field geologists followed from the simple fact that one of the major goals of Mars exploration is the understanding of early environments that could be habitable, and what could be more relevant than looking at the rocks that date from the earliest environments on Mars? This is the time-honored way that geologists go about understanding the history of a place. As an example, many natural history museums take the approach that we can look at environments through geologic time by examining the rocks and fossils that date from those ancient times. Beach sand that has been "fossilized," or compacted and cemented to a sandstone, shows that at some distant time in the past, there was a seashore where the sandstone is today. The age of that beach can be dated, usually by the fossils it contains if not by actual radiometric means. There are other things that rocks tell us, but this is one of the stories in the rocks that we can read. So, looking at ancient rocks on Mars is a good way to get at what was going on in the distant past on Mars. And given that one of the earliest things we see in the big-picture geology of Mars is lots of water and erosion from water, then the rocks formed in that era might have something to say about how habitable Mars may have been. But to get some sense of the bigger picture, the only way to do so is by moving horizontally over the surface. It is only by traversing over a "significant distance" that you sample the scenery and local terrain with your observations to start seeing the differences and similarities between different locations.

As for "doing geology" robotically, once you sit down and start sketching out the sequence of tasks involved, there is a very well-defined sequence of expert activities involved in accomplishing those activities. In the end we actually did design the mission and the use of rovers so that we could "do geology" on Mars, and it became part of the daily operational process of these rovers. The process of doing geology in the field is simple, but to my knowledge no field geologist really ever thought about it as a workflow. This is the power of interdisciplinary activities. What one group finds to be intuitive and based on repeated experience, another group with more systematic minds may find to be a simple pattern of activities.

For a geologist who is out in the field attempting to make sense of the geology, the process goes something like this and can be compared with the corresponding rover activity:

THE GEOLOGIST IN THE FIELD ...	THE ROVER ON MARS ...
1 scans the terrain to look for an outcrop that may hold information about the rocks. This might take a minute or two.	takes some images, especially a panorama. This usually takes one sol.
2 walks toward a selected outcrop. This might take several minutes.	does what we refer to as an approach drive. This would be the next sol.
3 looks at the outcrop up close and selects a particular part of the outcrop that can be analyzed. This might take a minute or two.	takes some images of the outcrop just after completing the previous approach drive. Another sol has passed.
4 either whacks off a hunk of the rock outcrop with a rock hammer to get a look at the clean interior free of weathering or maybe just looks at it up close. This might take a minute or two.	reaches out with one of the instruments on the arm. The rock abrasion tool would be the equivalent of the rock hammer. Another sol has passed.
5 looks at the rock chunk that was broken off to assess the minerals and other useful characteristics. If it is an important rock, a sample might be taken and placed in a bag for later analysis in the lab. This might take up to five minutes.	uses its tools to do the mineral and chemical analysis right there, so there is no need to take a sample. Another sol has passed.
6 repeats the process, a traverse is made to the next outcrop, and so on.	repeats the process, a traverse is made to the next outcrop, and so on.
7 if the goal is to map the layout of different units and their relative timing, records these observations on a map, particularly how one rock type contacts another, so that overlapping relations are recorded.	if the goal is to map the layout of different units and their relative timing, records these observations on a map, particularly how one rock type contacts another, so that overlapping relations are recorded.

It should be obvious from this that what a field geologist can do in a few minutes, it takes the rover many days to do. But in the end, we had a flow chart for how we "do geology," and the rovers could simply follow that routine. Armed with this hindsight, it is now hard to laugh at anything that someone from outside your particular specialty might say. They may just have a better perspective on your goals than you do. They may be addressing an element of what you do that you have taken for granted, but really have not clarified. This perhaps illustrates yet another example of what interdisciplinary work is all about and why it generally works so well and generates such useful new insights and results. Sometimes that other perspective allows users to open the hood on a convention and do some hot-rodding.

DAILY OPERATIONS

ASIDE FROM THE CHALLENGES OF OPERATING on the daily solar time of another planet, the challenge of commanding rovers through complex activities on an alien world requires an efficient organization of activities on Earth, not just the workflow for accomplishing activities, but the workflow for agreeing collectively on what has been done, what needs to be done, building a plan for doing it, and then getting it sent two hundred million kilometers or so to a rover that can do it, and doing this process repeatedly, every sol, for weeks, months, and years. For this we had a plan that had been developed on Earth prior to the mission. A big part of that plan was what became known as the daily Science Operations Working Group (SOWG), a long name for what was basically the science team group that reviewed what the rover had done the day before and got together to plan what the rover would do the next day.

The SOWG was the meeting in fact in which we made plans to be uplinked to each rover and it had a very organized structure with a rotating list of team members including chairperson, representatives for each of the instruments, representatives for several science "theme groups," a documentarian, a long-term planning lead, and a panel of mission engineers. Initially we all met in a room in Building 264 at JPL, designed specifically for this purpose, that consisted of a U-shaped arrangement of tables, a "back row" of engineers, and two large projection wall monitors used for displaying information and plans being discussed on each day. Early in the mission this room became known affectionately as the "Callas Palace" because it was such a lovely setup. The name referred to John Callas, the excellent JPL mission project manager who was responsible for the big job of organizing, managing the team, and providing all the tools and equipment required to get the job done.

Many of the team members had important roles during the SOWG. The SOWG chair led the meeting, the documentarian took notes that summarized what we had all discussed for future reference, the instrument team leads reported on the data and status of each instrument, the theme group leads proposed observations based on the desires of the science being done at a site, and the long-term planning lead kicked off the discussion with a brief presentation reviewing what was previously planned, what was downlinked, and what the strategic goals were at that point in the mission, along with other information that might help in the efforts to build the plan for the day. There was a SOWG meeting, usually daily, for each rover. Actually, it was a bit more complex than that, but this was the essential structure. And while it sounds complicated and somewhat unwieldy, we got good at it. Several thousand such meetings happened throughout the lifetime of both rovers. We got very good at it after many years of working together and often completed the process of "building a plan" in a half hour. As the mission progressed and we all had returned to our respective institutions, the meetings took place by telecon and internet, with team members at their home universities or institutions across the country. All this activity would continue, almost daily, for years, with team members also working on their regular faculty, teaching, research, and other tasks at their home institutions. Before the landings took place, we had several practice sessions with the whole process, called operational readiness tests or ORTs. including practice with operating a test rover on a virtual Mars in another building. This was to ensure that each of us knew our jobs and so that we could hit the ground running.

As both spacecraft approached Mars in late December 2003, it was time for everyone to migrate to JPL where we would operate as an on-site team for the first three months of the mission. Once again, I packed the car and drove out to Pasadena, just as I had done more than two decades before for Viking. The big day was at hand, and that day would be the first week in January 2004, with the landing of MERA or Spirit.

> But in the end, we had a flow chart for how we "do geology," and the rovers could simply follow that routine.

ACOMA GOES TO MARS

AS WITH MANY NASA MISSIONS THERE were plans for a variety of public awareness and educational outreach activities. One effort that was particularly fun for many of us was a small program named the Athena Student Interns Program, the name Athena referring to the instrument package name. This program was coordinated by an experienced NASA educator, Cassie Bowman. The goal of the project was to have a teacher and two student interns selected from high schools from across the nation to be present for two weeks during the mission and to work with a science team member in their state or region. So we were given the opportunity to put out a call locally to invite proposals from science teachers in our state, the winners to be invited to come out to JPL for a short experience with an active mission.

Jayne, as the geoscience/space science educator at the New Mexico Museum of Natural History and Science, helped me put out a call to local educators. NASA actually selected the final thirteen teams that were chosen nationwide; and we had a winner in New Mexico. In our case, it was teacher Joseph Aragon and his three selected students, Brandon (Jay) Herrera, Mark Vallejos, and alternate Henry (Hank) Vicenti, from Laguna-Acoma Jr./Sr. High School located on the Acoma and Laguna Pueblos of western New Mexico. This school serves grades seven through twelve, with a total student population of 377 that is an amazing 93 percent Native American and Hispanic, and is located on rural reservation lands.

Thus began a friendship that has lasted to this day. Joe Aragon was, at the time, a science and math teacher at the high school level and from Acoma Pueblo. Jay was, at the time, a high school junior from Laguna Pueblo. He has since earned his master's degree in engineering and has worked as staff engineer for Laguna Pueblo. Mark, a senior at the time, was from the nearby traditionally Hispanic mountain village of Seboyeta. Mark earned a master's degree in social work and works with underserved rural communities. Hank was also a senior and from Acoma Pueblo. He was appointed as a traditional Acoma Tribal Sheriff in 2019.

Acoma is the oldest continuously inhabited town in the United States, although there is some rivalry for that title between Acoma and the Hopi Mesas. Acoma traces its history back one thousand

BELOW: Joe Aragon, Mark Vallejos, and Jay Herrera in a Mars analog site near Acoma and Mesita Pueblos, New Mexico.

years and sits atop an impressive sandstone mesa in remote western New Mexico and like Laguna Pueblo, it is surrounded by reservation lands and several small and historic Hispanic communities, of which the small, colonial village of Seboyeta is one. Joe Aragon is an impressive guy, big and energetic with a sense of humor that served for all occasions, with degrees in physics and math, and M.Ed in education. He is a full member of the Acoma Pueblo, at one point serving as lieutenant governor of Acoma, among other official duties, and he is an adept and experienced educator. It was from Joe that I learned of the Native American view of many natural things, including the role of Coyote as both a trickster and a wise educator. He has a commanding presence and when he speaks you know you will be hearing something at once interesting, important, and possibly ironic, just like Coyote. Joe, a talented artist as well as educator, even drew up an Acoma-themed logo for his school's participation. It was appropriately referred to as the Spirit Rover.

From the start it was clear that Joe, Jay, and Mark had a leg up on all the other student intern groups in that their backyards were essentially analogs for Mars. Many of the concepts that we think of when we think of the Martian landscape were part of their daily lives: arid environments, lots of rocks, warm days, cold nights, and all manner of landscapes, including volcanism. In fact, as a lead-up to their participation, Jayne and I took Joe and his students on a series of short field trips to nearby locations to talk about field geology and the types of observations that we would be doing on Mars.

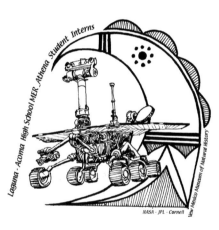

ABOVE: Acoma-styled logo showing the Spirit Rover with a framed heraldic of the Eagle Clan.

Each group of student interns would visit, and with guidance from their teacher, be a part of the science team for two weeks, and the groups would rotate through in sequence. As luck would have it Joe, Jay, and Mark were the first group, and so they were there on the day of the first landing. They got to experience the drama of that event and then went on to experience the first days of the mission, even participating in some of the first science activities as well as naming a couple of rocks. Jay named one rock Frybread because it looked like a small piece of that particularly tasty Native American treat laying on the surface.

Their time at JPL came to an end right after the landing of Opportunity, but our friendship has continued to this day. I believe it is safe to say that they were the hit of the entire program. The science team and engineers at JPL were extremely impressed by this team from rural New Mexico. Basically, they blew all the other high school teams (most from large cities and schools) out of the water! And they were from a small school in a remote and Martian-like corner of the New Mexico desert. They were naturals in exploring Mars. The team was able to attend the annual Lunar & Planetary Science Conference in Houston that year. They also presented their experiences to other nearby rural school groups and continued to work with us for the first few years of what became more than a decade of Mars and rover mission-related educational programs created by the New Mexico Museum of Natural History and Science.

As the only museum other than the Smithsonian Air and Space to have a direct connection to the mission, we took full advantage of the opportunity. We were given permission by JPL to build our own full-scale, detailed replica of the rover as part of a permanent Mars exhibit, and later added a special tenth anniversary exhibit. While I was living in Pasadena and working at JPL on Mars time, Jayne and colleagues, including Joe Aragon, wrote a rover resource and activity guide and created education programs that ultimately reached several million visitors and tens of thousands of teachers and students, and the general public. We learned some interesting lessons over the next few years, including the role of museums as trusted community science resources, and the importance of a planetary geology mission, exploring the surface of a new world, in exciting and educating people about basic science.

ARRIVAL AT MARS

SPIRIT—OR MERA, AS THE PROJECT nomenclature referred to it—was the first to arrive at Mars. On the night of the arrival the science team was sitting in a big room that would become the daily SOWG meeting room, the "Callas Palace," surrounded by monitors giving updates on the spacecraft position as it began the final long plunge toward Mars. Everyone was whiling away the time absorbed in personal tasks, including the ubiquitous tapping on the keyboards of their laptops, probably doing emails or maybe even taking care of other science business that we all get such as manuscript reviews or even revising our own manuscripts. You would almost think that nothing was happening watching everyone sitting at tables around the room absorbed in their

"Perhaps this was a day of firsts. The day one dies, of course, is a first in any life."

—DEAN KOONTZ, *Saint Odd*

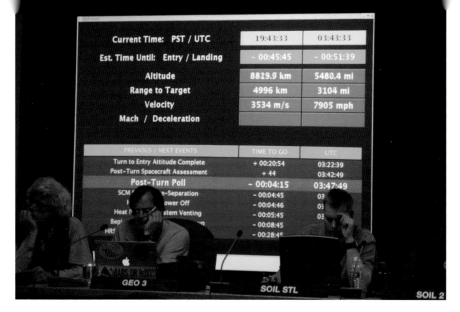

ABOVE: Members of the science team await the landing of Spirit in the uplink room. On the wall is a projection of spacecraft progress, at this point just before atmospheric entry.

laptops. But right behind them, projected on the large screen along the wall was an informational readout of spacecraft position, distance from Mars, and velocity. The distance from Mars was a blur of diminishing numbers and the velocity was a blur of increasing numbers. You really felt as though you were riding the spacecraft down, plunging toward your glory.

All human endeavors that might be described as grand and dramatic spectacles and that are repeated periodically develop their own traditions. Launches are one example, with the traditional countdowns and fiery blasts-offs that are the culmination of years of hard work and preparation. Landing on another planet has become another one. It is also a process filled with all sorts of sudden-death scenarios. Landing on another planet is a difficult event for the human brain to completely assimilate. There you are in the form of a spacecraft crossing interplanetary space, traveling at several thousand kilometers per hour, and you go from being this cosmic being with a grand view on a big scale in which planets are floating serenely in space to being a mortal squatting in the rocks and dirt of a planet's surface. And the transition happens in just the space of a few minutes. One minute you are in interplanetary space, where everything appears to move in stately slow motion, and the next few minutes you are hurtling in what seems like a fiery, falling death dive to a planet's surface. The heat shield you used to protect you during the fiery drop planetward is still cooling when you pop a 'chute, and it is hardly out when the ground races toward you and suddenly rockets roar into action, you jerk to a halt still way above the ground, and then you

Interplanetary travel may be something that is outside the everyday experience, but it has a lot in common with some everyday experiences.

are cut loose and drop, bounce, and roll. Other subsequent rover missions have talked about seven minutes of terror, but Spirit, and our human mission team, were the first to experience those minutes of silence.

It turned out that Spirit had been cut loose a little high and fast and at a slight angle that made it bounce and bounce and bounce and roll for what seemed like an eternity before it finally stopped, unfolded its lander petals, and communicated with us. And then, in a blink, suddenly you are sitting there looking at horizons like the ones you see every day as an Earthbound human. Except the landscape is totally alien, the sky is sand-colored, and the whole scene is a hundred million kilometers away from those you are used to seeing. Now with every blink and look, you are seeing what has never been seen before. You are on another world, an unexplored world. And you get to be the first person to explore it. This hasn't happened for a hundred or more years on Earth. It's a grand natural history expedition, but in this case to another planet, and it's the twenty-first century, not the nineteenth. It is a twenty-first-century natural history expedition to another planet.

Interplanetary travel may be something that is outside the everyday experience, but it has a lot in common with some everyday experiences. For example, I like to think that getting to Mars is a bit like the way your cat alights on a shelf after having jumped up from the floor. If you watch carefully you will notice that the cat arrives at the surface of the shelf in that precise moment of free fall at the top of its trajectory. From there, the cat literally floats down from the apex of the trajectory to the surface of the shelf under an acceleration or pull of less than one Earth gravity. It is very elegant and very precise; cats must have a good sense of the shape of a parabolic trajectory. It must be one of the enjoyable things about being a cat, and cats are always giving themselves a mental "oh yeah!" when they stick it just right, somewhat like the feeling we get when we "strip the net" of a basketball hoop from a long shot. But it is also very similar in many ways to interplanetary trips that put you on the surface of another planet.

For an interplanetary shot, Mars catches up with the spacecraft and the spacecraft starts arching in from its otherwise straight trajectory slanting out ahead of Mars. The spacecraft follows a path shaped somewhat like a strange, distorted, and warped parabolic trajectory. The final part of that trajectory, the part after the arch at the top, is just very short compared with the part arcing outward from Earth to Mars. From the spacecraft point of view it looks like you are topping out on the trajectory and slowly starting to fall back, but instead of falling toward the surface you launched from, now there is a new surface or gravity pulling you back. At this point you are now approaching Mars from the west as it catches up to you and catches you in its gravity. The descent begins obliquely into the atmosphere on a low-angle trajectory. From Mars's point of view there is this crazy spacecraft that is trying to shoot past it and would fire across in front of Mars's orbit and get splatted on "Mars's windshield," but it gets trapped by Mars's gravity field as it approaches and gets sucked down to the surface, which slides under it along Mars's orbit, so you come in at an oblique angle and slide into the atmosphere.

On that night we continued to sit collectively attempting to be productive at something while we nervously waited out the final hour, and one by one there was a closing of laptops and all eyes were staring at the screen. Then in the last several minutes the screen changed to an animation of the surface of Mars, as it would appear from Spirit coming in hot. Mars was doing a stately roll "under our feet" as we did that long, long slide into the atmosphere and flew over a cratered landscape somewhere far west of Gusev Crater. And as we watched, the motion was visibly slowing down from scrolling landscape to one that was rapidly approaching with a clear vertical component to the motion. There on the top of the screen was Gusev Crater, and it was coming up obliquely and fast. Soon the whole crater was nearly filling the screen, and the surface was rapidly approaching now.

Calls went out that we were into the dynamic phase of atmospheric entry. Then the chutes are out! Followed by interminable seconds, minutes? Then in quick succession, bridle out! Airbags deployed! Braking rockets! Bridle cut! First bounce, another, another. Motion stopped! First signal from the surface! We are alive on Mars! Bedlam erupted. We were at the end of the long road to Mars and we were on the surface and still alive. This was a major milestone that had happened only three times before, twice in 1976, and once again in 1997. Minutes pass, airbags retract, panels unfurl, and the first picture is returned. We were on Mars again.

After Spirit was landed and we had a day on Mars under our lander, we set about working on where exactly on the map we had landed. In those days we had some regional mosaics cobbled together from what at that time was the highest resolution data, the Mars Orbiter Camera, or MOC. We had a big wall-size mosaic "map" of the landing ellipse on the floor of Gusev Crater in Building 264 at JPL. The question for the team was, where did we just land? The engineering team was pretty sure we were inside the ellipse. But what was not clear was whether we landed near the center or somewhere uprange or downrange of the center. One of the reasons that additional uncertainty had crept in was the fact that there had been a global dust storm just before the two rovers arrived at Mars. It was a failed attempt by Coyote Mars no doubt to do a Mariner 9 number on the mission. Or perhaps it was just a little joke to keep us guessing. Mariner 9 of course arrived at Mars at the height of a global dust storm that veiled the planet in a uniform blanket thus frustrating initial attempts to get the mission fully underway. But in this case the dust storm was not an enormous one and took place a little earlier, so things had cleared up somewhat by the time Spirit arrived. However, there was still some dust in the atmosphere because it takes months for lofted dust to completely abate. The significance of that dust was its potential effect on the entry trajectory because dust increases the temperature of the atmosphere and expands it outward from the planet. Therefore, the narrow time frame of the sequence of events necessary to get to the surface must be tweaked. But how much and exactly when is a relative gamble based on limited modeling.

To pass the time right after the first landing and before we pinpointed our location on Mars, we had a little contest for the science and engineering teams in which everyone could make their best guess at the location of Spirit on the surface of Mars on this big MOC mosaic of the target area on the wall. The ellipse represented the outline of the area within which the lander had a high probability of arriving at the surface of Mars. The ellipse was oriented with its long axis slightly north of east and roughly eighty kilometers long and twelve kilometers wide. The area of probable landing is elliptical because it is really the projection of a circle of uncertainty obliquely into the surface of Mars. In other words, the ellipse is like the elliptical beam of a circular flashlight when directed at an angle to the ground. The size of the circle itself is based on all sorts of considerations about the ability to "hit" a target point on Mars after an interplanetary journey, including things like the fundamental uncertainty of pointing to a precise point on Mars given the uncertainty in surface coordinates. But really the uncertainty of controlling all the location and position information on a journey through interplanetary space and particularly the timing necessary to arrive at a certain point at a certain very precise time are some of the more important factors in the process. NASA has gotten better and better at shrinking the size of the area of the high probability of landing over the years, but when Spirit and Opportunity landed the high-probability area resulted in an ellipse about eighty kilometers long on the path of the descending lander, about twelve kilometers wide, and centered on the targeted location.

This was the "formal" area of uncertainty, and the engineers responsible for calculating these things would unofficially suggest that there was a very high probability of landing within an area about half the size of the formal ellipse. Perhaps it is similar to the situation when Scotty would say that the engines are close to 100 percent power, but in reality, he thinks they have another 40 percent and is holding that in case more power is really needed. Or at least that is the way many of us like to think it works. Later we would become so accustomed to that thought that when the engineering team stood by a statement that we could not do a thing that the science team wanted to do, it was disturbing. It was as though some fundamental rule of the game was not working for us.

On this large wall image map, each of us were to take marking pens and put a dot with a number next to it showing where we "guessed" was the final landing location. We then wrote the number and our name on a clipboard hanging from a string.

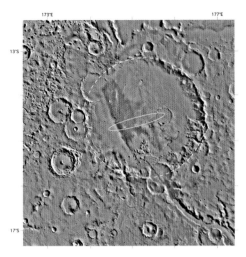

ABOVE: Viking image of Gusev Crater, more than 161 kilometers across. Shown in yellow is the prelanding ellipse indicating the probable area in which Spirit would land assuming a target near the ellipse center. The small star inside the ellipse indicates the final landing site location, which was ultimately a bit downrange from the center of the ellipse.

Once the official word came down regarding the location, the person who ended up being closest to the actual landing site would win.

I think a lot of dots on that map were the equivalent of blindfolded dart throwing. Dots were all over the image. Curiously there were lots of dots clustered uprange or short of the center. I am not certain what that was all about but there was probably a beer session buried under that particular batch of dots somewhere. But it seemed to me that you could increase your odds of getting closest to the actual landing point by "thinking like a lander." It was clear just from listening to the engineers that the big uncertainty was the fact that the atmosphere of Mars was a little warmer than modeled in the original landing targeting predictions. The reason it was warmer was because of a significant dust storm that had occurred just prior to landing. And, as I said before, dust lofted into the Martian atmosphere has the peculiar effect of warming the air. And for interplanetary spacecraft seeking to enter that atmosphere and land, the more important consequence was the fact that the atmosphere was somewhat expanded and extended farther out into space because of the additional warmth. This means that the top of the atmosphere was just a bit higher than the "nominal" models had used. As the spacecraft approached Mars, the first resistance by the atmosphere during the process of entry occurred just a little higher than originally assumed. So

the sixty-four-dollar question was whether that would slow the spacecraft down more and it would arrive short, or "uprange," of the target, or whether that would cause the spacecraft to "skip" through the upper atmosphere a bit such that it ultimately landed beyond, or "downrange," of the targeted landing site?

Trying to think through how this might all affect the trajectory of Spirit as it was coming in, it seemed to me that coming in at a shallow angle could be more like a flat pebble skipping along the surface of a pond. So I was in the "downrange" camp of guessing where the landing point was located. Next, I needed to turn that thinking into a point on the map. To do just that I stood "uprange" of the map and looked at the map obliquely until the ellipse outlined on it looked like a circle from my perspective. Remember that the actual targeting uncertainty was a circle in space, and like a circular flashlight beam aimed obliquely at a surface the circle was smeared out on the surface due to the oblique entry angle of the spacecraft. I looked at the ellipse from uprange until it looked like a circle, and then I pointed my arm and fingertip at the target in its center. Then to account for the warmer atmosphere I slightly raised my arm so that I was skimming, floating on the warmer, thicker atmosphere just a tad downrange of the target. To account for the slight delay in arrival resulting from the longer path, I moved the point a little to right. Then I followed that point to the surface of the map, and that is where I put my "dot." Amazingly enough, when all was said and done and the final landing site verdict came down, I was just a few hundred meters east of the true site and had won the contest.

TWO MISSIONS, TWO TEAMS

IMMEDIATELY FOLLOWING THE LANDING ON WHAT became known as sol 1 (or Mars day 1), Spirit began setting up shop and getting ready for its first small step for roverkind when it would roll off the lander and onto the surface and begin what was to become an epic journey on another planet.

On sol 1 for Spirit, the first look at our surroundings elicited mixed emotions depending on your professional science preferences. The scene before us was a rocky and dusty red plain. And the rocks were dark, angular, and clearly pitted in the way that basaltic rocks are pitted with gas bubbles or vesicles. So Spirit had landed in a basaltic plain, or in other words, a plain that was old lava flows instead of the lake bed sediments postulated for Gusev Crater. Coyote Mars had once again done the old Mars bait-and-switcheroo on us with the Gusev Crater landing site. Despite the collective wisdom of the selection process we had ended up in pretty much the same setting that had demoted a site on the south margin of Isidis basin that I had proposed in several landing site meetings, a site that was more than likely going to be lava plains rather than sediments washed out from nearby highlands, requiring, as we later were to find with the Gusev site, a long drive to reach the older rocks that we really wanted to investigate.

On the one hand, the site was not what we had anticipated in terms of the presence of deposits from what must have been an ancient lake in Gusev Crater. On the other hand, here we were, once again, at long last staring at a new scene from the surface of Mars. I was stunned because I was now a part of a new and grand geological exploration adventure. We had successfully made it to the surface. There were to be no massive disappointments at having such an exciting mission and opportunity to do field geology on Mars slip from our grasp.

The initial scenes at the Gusev Crater landing site included nearby dusty small craters that we at the time referred to as hollows just in case something else was going on and the depressions were not impact craters. It is common for certain types of lava flows to have surface depressions as part of the process of lava flow emplacement, so it was thought that just in case that was what these depressions were, we should use a generic nomenclature for depressions like the ones we were seeing even though they looked like eroded impact craters.

During those long nights on Earth when we were getting accustomed to the off-Earth (Mars-time) work schedule, many people on the team were a bit sleepy, so the nearest, shallow, dust-filled hollow became known as Sleepy Hollow. On the floor of

Sleepy Hollow was a clear impression of one of the last bounce marks from the airbags that were used in the arrival at the surface. In fact, you could see a series of disturbed roughly circular patches in the soil that were bounce marks marching across the surface leading up to the final resting spot. Rocks were everywhere, but in an abundance that science team member Matt Golombek, JPL, would assess in the first mission reports as being pretty much the same as predicted from orbital analyses. But in the distance to the east-southeast were a series of "mountains," hills actually, that we later named the Columbia Hills in honor of the lost *Columbia* space shuttle of a few years before, with each summit named after members of the *Columbia's* crew. The hills were beckoning, but seemed impossibly far away, at least 2.5 kilometers. For a mission that was only planned to rove six hundred meters and last three months, those hills were considered out of reach by any stretch of the likely mission timeline. Little did we know at the time that our assumptions were to be proven wrong on that score, and Spirit would become the first mountain climber on Mars as it traversed the Columbia Hills. And it would go on from there for many years more.

From our first close looks at the surface from the lander it was pretty clear that these were volcanic rocks, basaltic rocks from lava flows that must have flooded the crater floor, forever obscuring any lake bed sediments from access and the source of all of the disappointments about the lack of obvious sediments. Later during one of our many team science meetings fellow science team member Dave Des Marais, NASA Ames Research Center, was to lament in what I often refer to as an "infamous comment" that Spirit was trapped in a "basaltic prison," in reference to the fact that Opportunity over on the other side of the planet was enjoying a setting consisting of obvious sedimentary rocks and all manner of chemical stories about the past presence of water. Dave and I became good friends, but we had to disagree on this. Dave was coming from the perspective of water and possible biology, and I as a volcanologist. I thought that the Spirit landing site was lovely and somewhat more scenic, but liked the fact that it was a primary volcanic surface. Unlike the previous landing sites of Viking Landers and Pathfinder, where rocks had been washed in and jumbled, the basaltic rocks of Gusev Crater were still in the place where they had formed and by

Part of the first panorama showing the view south from Spirit's landing site. Bounce marks from the landing airbags are visible on the dusty small crater floor on the left.

comparison had been only partially beaten up over the billions of years since they flowed across the surface. All those volcanic rocks were just sitting there waiting for us to decipher an important part of the volcanic history of Mars and see what lava flows on Mars really had to offer. But the MER project was predicated on the study of sediments and everybody "knew" that sediments could preserve the ancient environments of Mars. But hard rocks preserve evidence too, just in a different way.

Spirit was a drama queen afflicted with all manner of events over the course of its explorations, a characteristic presaged by its first drama, from the start.

The journey of exploration began with the first step onto the surface on sol 13 after a lengthy checkout of the systems and a decision by a sleepy and anxious team, living and working on Mars time, on where to begin. We immediately made use of our shiny new instruments for a quick look at the soil near the lander. One day over two weeks from landing we rotated toward the east and moved toward a big rock, where we would determine what these rocks were that were strewn all about. There was a nice flat-faced rock just a few meters away that had been given the unlikely name Adirondack that appeared just right for trying out our tools for examining rocks.

We arrived and moved into position with the rock centered in the robotic arm workspace, deployed the arm, and set up to examine the rock. A series of tasks played out over several sols in which we sent commands and waited for the results at the end of the day. The next sol we would send commands to begin, but when we came in the next sol, "there was a problem, Pasadena." Spirit was not talking to us. This was the beginning of the so-called sol 18 anomaly. During this period, it was realized that Spirit was continuously waking up and shutting down, unable to receive Earthly commands and consuming valuable battery power doing so. Spirit was having a nervous breakdown and no one could immediately determine why.

In the end, the engineering team was able to puzzle out the principal pieces of the mystery, which were related to a problem that many of us have seen with our personal computers. It was trying to do too many tasks with too little memory. And part of the problem was that the flash memory was filled with data from observations taken up to this point and all that data had as yet not been downlinked and deleted from memory. Both rovers were capable of accumulating more data bits than could be sent on the downlinks to Earth. So data accumulated unless managed. Later, part of the job of long-term planning lead, an operations role that I, along with several others, held throughout the mission, was to summarize the state of the onboard flash memory so that observations could be throttled according to how much data we could easily return. This data management task became routine thereafter in response to this scenario, the first of Spirit's "drama queen" moments, as Steve Squyres had later referred to them. Thus we learned something valuable from the anomaly. But the attention of the science was necessarily diverted at the beginning of the anomaly for a sol or two because this was right at the time that Opportunity was on schedule for landing over on the other side of the planet in Meridiani Planum.

Spirit was a drama queen afflicted with all manner of events over the course of its explorations, a characteristic presaged by its first drama, from the start. Opportunity, the older sibling but the second to actually arrive on Mars, appeared to have all the luck, and by comparison, if not in detail, it gracefully wandered from discovery to discovery. Not only was Opportunity sent to a surefire successful site where we knew there would be something relevant to the quest for evidence of ancient water, with widespread evidence of hematite but it also had a relatively flat "parking lot" for a landing terrain. There were no rocks, just a smooth, dark, and pebbly-appearing plain.

But to sweeten the deal, Opportunity did an interplanetary hole in one and landed in a small crater with gently sloping walls. This was a shallow crater and not so steep-walled or deep that there was any concern about being trapped forever. To add to its luck, if more were needed, it was clear from the first panorama that there were nice layered outcrops nearby in the crater wall. This was a site that had been selected, of course, for the abundant orbital evidence for hematite, a mineral that tends to occur in the presence of water, and the whole of Meridiani Planum appeared from orbital mapping to be one giant deposit of some type of sediment that had originally buried the ancient landscape on this part of Mars. So layers meant sediments and nearby layers meant plenty of outcrops to exploit for investigating the origin of those sediments. And instead of rocks strewn all about as we had seen at Spirit's landing site, the terrain around Opportunity was oddly dark and pebbly in texture, and not a rock was to be seen outside the outcrops nearby. And later, after Opportunity climbed the rim of the crater, the view of the surroundings was stunningly flat with the same dark smooth surface extending out to the horizon.

Opportunity had landed in the type of geologic setting that the mission needed to accomplish its goals of studying past environments bearing evidence for past water, in contrast to the stark volcanic terrain of Spirit's site at Gusev Crater. A curious result of the perspective without an obvious scale had made the outcrops on the inner wall of Eagle Crater where Opportunity had arrived appear to be giant cliffs, but in reality, they were mere centimeters high. This was a boon for analysis because Opportunity could easily approach and inspect them with the instruments on its arm. The

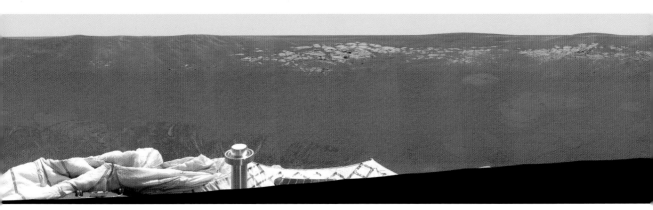

ABOVE: Part of the first panorama showing the view from Opportunity's landing site inside Eagle Crater. Outcrops of the Meridiani Planum sulfate-cemented sandstone are exposed inside the crater wall. Bounce marks from the landing airbags are visible on the dusty crater floor on the right. BELOW: Microscopic Imager view of the "blueberries" or hematite concretions on the surface of Meridiani Planum. Image is about 3.5 centimeters across.

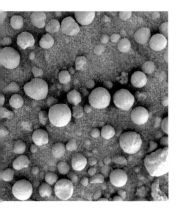

first examination revealed many of the geologic features that we had hoped for. Inspection with the chemical tools showed that the rocks contained minerals that form in low-temperature settings in dry lake beds. And examination with the Microscopic Imager, akin to a geologist's field hand lens, further revealed the presence of textures including cross-bedding structures typical of water-laid sediments. While the environments on Earth in which those minerals occur are highly saline and caustic, it is known that some forms of life can exist and thrive in those hostile places because in the end they are places with water; not much water, but possibly enough for hardy microorganisms of the time when the deposits were laid down.

Then there were the so-called blueberries. These were little spherules that were everywhere on the surface like small ball bearings, each a few millimeters in diameter. They were nicknamed "blueberries" simply because they were blue in initial color-stretched images, a thing that happens when the generally rusty light brown of a color image is contrast stretched such that reds appear redder and less red things appear blue. But the name stuck and everyone was fascinated with them. When analyzed in a fortuitous little cluster, they were determined to be iron-rich, but an oxidized type of iron known as hematite. Many theories have been proposed for their origin and ubiquitous occurrence, too many to review here, but a popular one supported by the science team in the light of the sedimentary setting was that they were hematite concretions formed in rocks that had been sitting in stagnant groundwater. They were concentrated on the surface presumably because they were uniformly distributed in the sulfate-cemented sandstones of Meridiani Planum, and subsequent erosion of the less resistant rock mass in which they occurred left them on the current erosion surface.

Similar concretions are common in sandstones on Earth in the Southwest and are a popular thing for rockhounds to collect. One could go on and on about how they form and what they mean. In fact, while the existence of concretions has been well known on Earth, it had been considered an interesting little phenomena that was associated with thick sandstones but not of particularly special interest . . . until we found what appeared to be concretions on Meridiani Planum. Suddenly hematite concretions were all the rage all over again in Earth geological circles. Like all analogs to things we see in more exotic locations, such as other planets, there was suddenly some value in studying an interesting but not particularly useful geologic curiosity. But in this case, if they were concretions, they pretty well implied that the sandstones of Meridiani Planum had at one time long ago been steeped in groundwater. Yet another piece of evidence that water was fairly prevalent even after the sandstone was initially laid down.

> **Suddenly hematite concretions were all the rage all over again in Earth geological circles.**

The focus on sedimentary environments was, in my opinion, single-minded. There are other ways for ancient environments to be preserved in the rock record, and geologists have used them often in research on Earth's rocks. For Spirit, later it would be shown that there was plenty of evidence of past water at the Gusev site, just not the picture-perfect layer-cake-geology kind with which non-geologists are more familiar. As Spirit's traverse unfolded, it would find rocks with veins and cracks filled with fluid-mobilized minerals and salt-crusted soils. And further still, it would encounter many rocks that had been weathered in a way that happens when rocks are soaked in water for long periods and exteriors are hardened while the interiors crumble away. For some reason that escapes me to this day, no one appeared as excited about these water-soaked rocks and many examples were left undisturbed in the Martian dust in the press to find layered rocks of the sedimentary kind. Like the "blueberries" of Opportunity's site, these water-corroded rocks were evidence of soaking water, but they were not seen as being as significant. Perhaps it was the mineral-, chemical-, and sedimentary-centric focus of the bulk of the science team. Perhaps it was Meridiani Planum sediment envy. There are other possible reasons, but in the end perhaps it says that any exploratory team should probably have a diverse array of experience in many different types of geology and equally loud voices throughout. But exploration is a human endeavor and we carry our foibles with us even on grand quests.

The moral of the story as regards landing site selection became obvious. First, pick something that you know must have a high probability of being true and then go prove it. The problem is, how do you qualify "must be true?" We learned that you need some observation or measurement that does not require a complex hypothesis but is a raw observation of a fact. For Meridiani Planum, the presence of the mineral hematite, as observed by remote sensing, was a smoking gun for past water. Second, and perhaps equally important, pick something that is universally understood by non-geologists who may be in positions to judge the success of your tests. For Gusev Crater it was a circumstantial kind of evidence of past water, a likely lake bed. For Opportunity, it was an in-your-face kind of evidence. Finally, the reality is that anywhere you go on a new world like Mars there's something new to be learned, provided expectations are set aside.

FIRST FIELD GEOLOGY ON MARS: THE INVISIBLE LANDSCAPE

WHEN I AM GIVING SHORT PRESENTATIONS to the public and students about Mars, a frequent question that comes up is what career preparation I did to become a planetary geologist. The question is always somewhat awkward because there is no simple answer that fits everyone. The answer is complex and really just depends on what fascinates you and what you are hoping to accomplish. And for most scientists, the preparation is a somewhat random walk amid personal interests that collectively lead to the end result.

I always had an interest in strange landscapes and what we could learn from them. When I was mapping volcanic fields in the Southwest, the geology was a horizontal geology. It was a geology not of eroded cross sections and road cuts, but a horizontal geology, with overlapping lava flow units. Figuring out the sequence of geologic units in my field research was not something that was measured in a series of layers, but something that was built from overlapping relationships. In this respect my field geologic mapping was always very Mars-like. It turned out that the field geologic mapping I did was in a terrain in which the landscape itself provided clues to the geology. In the end that type of study was perfectly aligned with what takes place with a rover on another planet.

There are two types of geology on Earth. One is hidden in the trees and eroded landscape and must be discovered through carefully looking at every outcrop, what few there may be. The shape of the land itself has little to do with the geology, except for the situation where differences in the hardness of rocks manifest itself here and there. But the shapes have nothing to do with how the rocks formed. The other type of geology is the one in which the geology is fully exposed, or if it is young enough, it is uneroded, and the shape of the land itself is part of the geology. The differences are illustrated best by an old eroded outcrop of granite and an uneroded lava flow. Both form outcrops that you can walk up to and touch and look at the minerals and determine the rock type. But one, granite, originates deep in the crust and is only exposed by erosion. This means that the outcrop itself is sculpted by erosion at or near the surface and you cannot use the shape of the outcrop to say anything about the rock except how it erodes. The other, lava flows, is formed at the surface. This means that the outcrop is close to the original shape it had when the eruption ceased, and the shape of the outcrop can actually tell you something about the events creating the flow. This second type was my habitual geologic field mapping environment.

So when we first began to talk about sending robotic field geologists to Mars, I had a pretty good idea how that might play out. Flat terrains that were favorable for landing or roving, "mobility" in rover engineering speak, were not going to reveal stratigraphy, that is, the ways rock units stacked with time. The best way to understand the stratigraphy was by continuous mapping along the traverse. This field geologic perspective on geology gave me a somewhat different approach to the mission than many of those concerned with the results of specific instruments. To me, the terrain itself was a piece of the geologic puzzle and any efforts that increased the observations of the terrain increased my understanding of how Mars geology worked.

"The only way to understand a land is to walk it. The only way to drink in its real meaning is to keep it firmly beneath one's feet . . . Only the walker can form the wider view."

—SINCLAIR MCKAY, *Ramble On: The Story of Our Love for Walking Britain*

EIGHT

One World, Two Expeditions

"Of two sisters
one is always the watcher,
one the dancer."

—LOUISE GLÜCK, *Descending Figure*

ONCE THE SPIRIT AND OPPORTUNITY ROVERS were safely on the ground and in very different settings, a new phase of science team activity commenced in which we adjusted to this new world of Mars exploration. There were roughly forty of us on the science team and a good half of the team rushed to the other side of the planet to participate in the perceived gold mine of sediments. The other half was happy with the scenic volcanic setting over at Spirit's Gusev Crater floor site. Part of the reason for this divergence of allegiance has to do with the basic schism of interests in traditional geology between igneous and sedimentary geology, also known as hard-rock versus soft-rock geology, which is a traditional way that academic geology departments focus on different specialties.

But the division of the team into two groups was also inevitable. The time difference between the local Mars times at the two landing sites and the difference in geology between the two sites simply made for a natural selection process. Regarding the time difference between the sites, the rovers had landed on opposite sides of the planet. Opportunity landed a couple of degrees west of longitude zero, while Spirit landed a few degrees west of longitude 180. This meant that when one rover was waking up and starting its day on Mars, the other rover was going to sleep on the other side of the planet. In practical terms this meant that the team working with one rover was starting their day when the team on the other rover was wrapping things up for that day. We were separated by close to twelve Earth hours in our work schedule. Many a Mars morning during the early mission days I would be marching from the parking lot at JPL toward Building 264 and encounter a member of the MERB team walking the other way at the end of their day, and we would each say something like, "Hey, you're going the wrong way!" The twelve-hour split between two sides of the planet also had the effect of suddenly reducing the number of team members that you worked with by half. Each rover had its own floor in Building 264 with identical facilities, except the walls and office furniture were color coded, red being Spirit and blue being Opportunity. Because of the time differences and different floors we infrequently saw the team members on the other side of the planet except during regular all-hands meetings.

There were a few people who attempted to dip into both missions on a regular basis with the corresponding consequences for sleep deprivation. Not me, I stuck with Spirit because I thought that the soft-rock geology over on Opportunity's site was boring. Whenever I wandered up to the fifth floor to see what was happening with MERB, my eyes would glaze over with all the fiddling with sedimentary outcrops, sandstones, and low-temperature sedimentary geochemistry. Then I would quickly hustle back down to the fourth floor and breathe a sigh of relief that we had "real" rocks and "real" geology in Gusev Crater where there were so many geological things to explore. Later on, of course, I found the Opportunity site to be just as exciting in different ways.

THE TWENTY-FIRST-CENTURY NATURAL HISTORY EXPEDITION TO MARS

TRAVERSING AND EXPLORING A NEW WORLD on the ground for the first time is simply comparable in many ways to our experience on Earth during the exploration of new lands and continents. As both rover missions began to progress, the missions of the two rovers were about to become a modern expedition of exploration to another world that is reminiscent of the epic expeditions of natural history exploration that had played out over the course of the previous centuries on Earth. There are probably just a few moments in human history when a small group of humans stood on the margins of a vast new world, and it is no stretch of the romantic imagination that the arrival of two rovers on the surface of another planet was surely one of them.

The first significant drives of more than a few tens of meters with Spirit and Opportunity were some of the rarest moments in human history, the first steps into a new land akin to the first steps onto a new continent during those first explorations of the fifteenth century in the "new world." These initial drives on Mars were the first to cover distances beyond the immediate terrain surveyed in lander panoramas. There can never, ever again be another first drive on Mars. There will be first drives for another rover, or for humans on Mars, or other missions in general. But never again will there be a first time on Mars for all time. With those first few drives we became travelers on a new world where we began to resolve what the new world was going to really look like. It was similar to taking small steps from the shore of a new continent while looking around totally struck by the newness of this world. Everything was something we had not seen before. What strange curiosity might we encounter at the end of the next drive? What would the view be like at the end of the next drive? Coming in every morning after a downlink to see where we were and what new wonders had been revealed became a gift waiting to be unwrapped every day. Anything was possible, because Mars is a cold, dry place that has existed for billions of years. And nobody had been there before us to see the things that were there to be seen.

Naming things that we saw became a favorite treat and we used pre-agreed themes for choosing those names. On sol 206, as we were approaching a somewhat spear point-shaped outcrop we were using a naming theme of paleo Native American names for rocks and investigation targets. It was here that we acquired the Cahokia panorama and where Spirit would be doing some important observations, So I had an idea. Not to be outdone with the use of paleo Native American names, it occurred to me that the name Clovis would be appropriate, which of course refers to the Clovis point, a particular style of spear point discovered many years ago that was dated as one of the earliest paleo-American artifacts and at that time was a remarkable demonstration of the early presence of Native Americans on the North American continent. Of course, the discovery site is in New Mexico and it was named after the nearby town of Clovis, New Mexico. So I had the idea that Clovis would not only be entirely within the theme, but it also just so happened to be the name of a town in New Mexico. People all over the world typically loved that sort of thing when a local name was used for naming something on Mars. Later when the local newspaper in Clovis reported the use of the name in a short article, the mayor of Clovis was quoted by the reporter as saying, "We are always pleased when the name Clovis is used in a constructive manner." I was not sure what to make of that, but it sounded like there was some back history there, and I was not certain I wanted to know about it.

After more than fourteen years of exploration, I was able to use more names from New Mexico. Opportunity's final campaign was a drive down a sinuous valley on the rim of Endeavour, a valley that we named Perseverance Valley for the perseverance that had allowed us to get there fourteen and a half years after landing. We spent a lot of time at one of the last team meetings debating whether to call it an arroyo, a gulch, a gully, or a dale, or several other suggestions that I do not recall. Finally, we decided to call it a valley simply because that is a fairly nongenetic term, whereas the other options all denoted some action by water. The purpose for avoiding water-related terms was again rooted in a desire to avoid making any interpretation about its origin, much as we had done by using "hollows" instead of "craters" back in the early days of the mission. We were there to determine what the valley's origin was and names have a way of fixating the thoughts and swaying minds.

Once we had decided what to call the valley, we needed a naming theme for observation and outcrops in the Perseverance Valley campaign. Several were pitched, but I suggested a Spanish history theme using the names of places along El Camino Real de Tierra Adentro. Throughout the mission it had occurred to me that, with a few exceptions, the themes tended to be rather standard collective history names, including names and places that came up in the history of the standard American culture, specifically the ones in American history lessons that tend to be rather Anglo-centric. There were a few names related to the history of Native Americans, African Americans, and Asian Americans . . . not as many as there should have been, but a few. However, names and places associated with the great exploration of the Southwest by the Spanish were absent, as they commonly are in our country's history books. Whenever a Spanish history theme came up it was passed over. I had a feeling that the reluctance had to do with the centuries-long reach of a thing called the "Black Legend" spawned by England during its wars with the superpower of the time, Spain. That in short had given the Spanish a particularly bad rap in the eyes of most of the English-speaking countries that continues to this day. In the end it was agreed to go with my suggested theme of place names along El Camino Real de Tierra Adentro, the great trade and exploration route from Mexico into the southwest United States.

Naming things that we saw became a favorite treat and we used pre-agreed naming themes for choosing those names.

There are many places along that fabled road and many of them are along the Rio Grande in New Mexico. This meant that we named things after Santa Fe, Albuquerque, and Socorro, and other place names that existed during the active lifetime of El Camino Real de Tierra Adentro. I had succeeded once again in getting some place names on Mars named after places in New Mexico. And it was a bit of fun listening to everyone try to properly pronounce Bernalillo. It was a Spanish pronunciation lesson of sorts for all. Unfortunately, one of the last outcrops we visited with Opportunity was named after a place along El Camino known as Jornada del Muerto. One must be careful with names as they do seem to have a power of suggestion, and the Fates and Coyote Mars might have been listening, or so it appeared. Opportunity's date with fatal destiny occurred not long after visiting that unfortunately named outcrop.

With all these names flying around and decisions and discussions being made there was at least one naming "accident." Of course most of these names are completely unofficial, so they live on with a few exceptions only in the science papers, archives of the mission, or legends of exploratory travels on another planet. It is with that caveat that I sheepishly must mention one example that stands out for me personally. As Spirit ascended the slopes of the Columbia Hills around sol 400 there were decisions to be made regarding the traverse route. One prominent feature that caught my eye in the MRO/HiRISE orbital image of the area was a knob overlooking a deep valley where I felt that not only would there be a likely bedrock outcrop amid all the downslope rubble, but a good lookout overview of a valley in the side of the Columbia Hills. I believe that I had marked it on a planning map that I had assembled for purposes of discussing possible routes and emphasized that I thought it would be a good place to get a view. During the course of a science and engineering team huddle to discuss routes, we were going back and forth, and Steve Squyres finally concluded, "I think we should go to Larry's lookout." Immediately the engineers were referring to the target outcrop location as "Larry's Lookout," and the name stuck. So, we went on to spend nearly one hundred sols plinking on the outcrops at Larry's Lookout and the name found its way into many mission documents. It is an informal name of course. Maybe someday the name will be official, but if it does become official I will not be around to enjoy it, given that by the rules of the International Astronomical Union nomenclature committee, surface features on other planets cannot be officially named after living persons.

WORKING WITH ENGINEERS

THIS WAS ONE OF THE RARE NASA missions in which the engineers and the scientists got to work literally side by side—and got to understand and know each other better. And I think that made this a much better and more successful mission. Things were bound to rub off on both sides of the room. Or I should say from the front to the back of the room. In the daily SOWG uplink and planning meetings the engineering team always sat at the back of the room with their monitors and made suggestions, depositions, and estimates of various spacecraft capabilities. The object of the game was for the scientists to want to do something and the engineers to find out why we couldn't or, to make us feel bad about further pressing the question, why we shouldn't do that thing. Or so it sometimes seemed to us scientists. After a while we both got to where we could guess when the science team would want to do something and where the engineers might tap our shoulders with a word of caution. So we got good at guessing what the other side of the room was going to do or say about any given plan, or we liked to think that. Actually, because the engineers had numbers and facts in front of them about spacecraft health and capability, and the science team had science questions and goals in front of them about vague things like "we need to test this hypothesis by touching that rock," it felt like a little gift sometimes when they surprised us and said okay.

There is another perspective on this daily interaction with engineers that I believe made all of us better scientists, and it actually increased understanding and respect among us all in both groups. It turns out that engineers do a version of the scientific method all the time, or at least when they are designing something. They take measurements, analyze what their measurements mean, and then try to build something that makes use of that understanding successfully and in a known range of reliability. If they get the interpretation—and the measurements—right, then the final results are a thing that they "engineered" correctly and works exactly the way it should work or needs to work. The science version of that is observation, analysis of the observations, developing multiple hypotheses for explaining how the observed thing came to be, and testing the hypotheses. The last hypothesis standing is assumed to be the "correct interpretation."

A ROAD TRIP WITHOUT ROADS

IT IS A RARE THING EVEN on Earth to go off-roading and actually go off a road or on a long-abandoned trail that passes for one. It is a dangerous business because every inch forward may be over a deep hole or a sharp stick. You can get stuck irrevocably or break down far from easy extraction. The Mars rovers were off-roading in the true sense of the words. They embarked on the first modern attempt to drive beyond the immediate landing area on an alien planet, a place where there were no roads, and to do so without mishap in a place dustier than anywhere on Earth and colder than Antarctica, and all that using just solar power in a fragile spacecraft with wheels.

The landscapes of Mars are literally a new wilderness. It is all the more remarkable that Spirit and Opportunity ultimately ended up out-living and out-traveling their design lifetimes and travel plans by considerable margins. The magnitude of these journeys was remarkable because the original mission plans assumed a much shorter mission duration and travel distances. One of the requirements for "mission success" before launch was for both rovers to drive at least six hundred meters and

> "I will not follow where the path may lead, but I will go where there is no path, and I will leave a trail."
>
> —MURIEL STRODE, *"Wind-Wafted Wild Flowers"*

last ninety sols. The accomplishments in that regard were greatly exceeded. Spirit traversed 7.7 kilometers and survived six years in an incredibly rocky and hilly terrain, while Opportunity, ever the lucky one, traveled 45.2 kilometers in its parking lot of smooth plains, breaking the distance record for any vehicle ever driven on the surface of another world. And it lived for fourteen and a half years.

The lifetimes of both rovers were set in theory by the fact that they were solar-powered. Previous experience with dust accumulation rates on landers told us that after a few months the dust that would collect on the solar panels of Spirit and Opportunity would cover the solar panels too much to sustain the necessary power requirements for continued operation. Operation included such things as having enough power to drive, operate the instruments, and send radio signals. But there were other power requirements. In particular, the cold environment at Mars was brutal, and heaters were necessary internally at night to keep the electronics from experiencing circuit-snapping cold. Just the incredible cold followed by daytime warming cycles could be enough to tweak some vital component, particularly an electronic part or circuit board. The cold was so great that small heaters were necessary for motors and joints prior to any activity after the long, cold Martian nights, even though both rovers were close to the Martian equator.

During the early days of the mission, because of the many hazards and unknowns of staying alive on Mars, Steve Squyres continually reminded the team of the possibility of a short mission, particularly when the team began to show any tendency to dither in one spot just because there were many questions to be resolved there. He often remarked that both rovers landed with a fatal illness: reliance on solar power that should, according to predictions about dust buildup on the solar panels, lead to not survivable low-power conditions in a few months. And further, in the way complex hardware had of failing, these rovers were incredibly complex and anything could break at any moment. The mental picture he advocated was that there was a sniper on a hill somewhere, and that sooner or later this sniper with no warning would pick us off. There were other less tangible hazards. It was predicted from years of practice with remote rover operation previously on Earth that there would be days

> "Depend on it sir, when a man knows he is to be hanged in a fortnight, it concentrates the mind wonderfully."
>
> —SAMUEL JOHNSON, *Boswell's Life of Samuel Johnson*

---

lost, perhaps one out of three or more, because of downlink or commanding glitches. So the true cumulative number of sols for actual science would likely be way less than the actual days on the surface. While this factor was a real one, with practice over the coming years the number of such days was reduced dramatically until in the final years such events were so rare as to practically cause irritation that such a thing would deign to occur when they happened.

In the face of all these uncertainties of actual science activity on a distant planet, Steve felt that only by pushing on could we hope to amass the diversity of observations that would lead to a bigger and more exciting picture. This meant that there was a constant yin and yang about the timetable for activity at any stop where things looked interesting. Should we take a few more sols to complete a thorough investigation, or should we decamp after a reasonable cursory examination of a rock or outcrop in the hopes that there would be more exciting things to come? It was either a curse or a blessing depending on how much a given site had to offer in one's own specialty. This was to play out repeatedly over the course of the mission even when it was clear that the power situation and other initially feared factors were less worrisome. The sniper was still sitting on a hill somewhere and the rovers were the targets.

In this conflicting climate of the need to do thorough jobs of observation at the places deemed of value yet the need to continue moving, the rovers nonetheless were able to accomplish considerable traverses. Spirit's traverse wound its way from the landing site to a nearby impact crater, and from there across 2.5 kilometers of broken basaltic terrain to a series of small hills, the Columbia Hills. Eventually Spirit summited the Columbia Hills, drove down the other side, and arrived in a valley, on the floor of which was a curious circular patch of brighter outcrops named Home Plate for its resemblance to the five-point slab at one corner of the diamond shape of a baseball field. The idea of driving to the Columbia Hills was initially a dream. From the orbital images it was clear that the hills were older than the plains because the lava flows covering the crater floor completely surrounded them and were therefore much younger. They beckoned from across the rocky plains as the best hope for seeing something from the deep past and the geological time when the Gusev Crater floor was perhaps a lake. Even if nothing of the lake history was present, at least it was an opportunity to see rocks from the distant past. However, many on the team felt that the idea that the mission could last long enough to make the journey to the hills was unlikely.

During an end-of-sol discussion as Spirit was edging past the originally planned mission lifetime of ninety sols, a group was standing around a table in the science operations room looking at an orbital mosaic from the Mars Global Surveyor MOC on one side of the table and the rover panorama on the other side. So I wandered over to see what the impromptu discussion was all about, a thing that was necessary on many occasions in order to be in on anything important that was going down. They had started a discussion about beginning to consider what was the best science use of the rover in the coming days of the mission, and were discussing the crazy hope of traveling beyond the landing site and the small Bonneville Crater where it was located at that time to other more important science efforts. Maybe we could travel across the plains to those far hills, the Columbia Hills, where older rocks were likely. Gentry Lee, a colorful fellow with a permanently attached baseball hat and a long history of advising JPL projects as an intermediary with NASA headquarters, was advocating that we go for it. He went on further to say, "I predict that in a year Spirit will be perched on the top of those hills." We all glanced at the panorama with

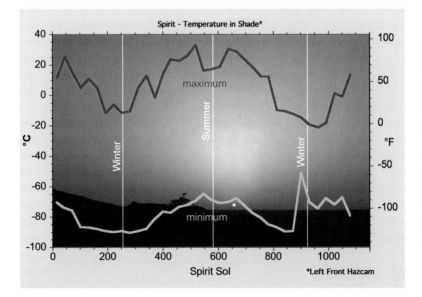

LEFT: Graph illustrating the daily maximum (high) temperatures in red and the daily minimum (low) temperatures in blue over the first eleven hundred sols on the Spirit rover front Hazcam cameras. Temperatures in the summer are moderate, but even in the summer the nighttime temperatures are similar to an Antarctic winter. Every night the temperatures fall nearly one hundred degrees, summer or winter.

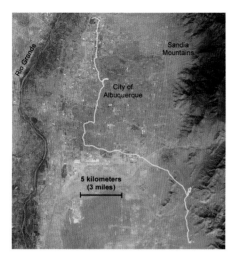

ABOVE: The length of Opportunity's final total traverse, shown in yellow on a map of a moderate-size city for comparison.

the Columbia Hills so far away across the rocky plains of the Gusev Crater floor and thought he was crazy. Maybe, just maybe we could get to the base if we were beyond outrageously lucky. But the top of the mountain? Good luck with that. Little did we know that within the next year he would be proven correct. I do not think anyone at that moment would have bet a nickel in favor of that outcome. It was a crazy idea, at least to us in our ninety-sol-circumscribed lifetime at the moment.

Meanwhile in the early days over at Opportunity, once it left the nest in Eagle Crater and after a thorough—and to many of us an excruciatingly protracted—investigation of the outcrops inside the crater, it began a journey that would end more than forty-five kilometers and fourteen and a half years later. The final odometry for the journey was akin to a circuitous drive in a golf cart across a medium-size city from north to south. Because the Meridiani site was so flat, the Opportunity science campaign took advantage of the scattered small-impact craters to access the deeper layers of the vast plain of sediment that was its home. The journey took it eastward to Endurance Crater for its first excursion inside a crater after a lengthy review of the mobility issues that would be encountered and an examination of a naturally exposed section of the layers in the crater wall. It emerged from Endurance Crater and continued southward on an overland journey to the even larger Victoria Crater. From there it wended its way through windblown ripples that proved more treacherous than initially thought, and onward to the rim of the largest impact crater, Endeavour Crater, some twenty-two kilometers in diameter. After arrival it drove an additional ten kilometers along the rim, exploring even older rocks exposed like an island in a sea of younger sulfate-rich sands that made up Meridiani Planum.

The idea of actually driving southward toward Victoria Crater after emerging from Endurance Crater unfolded in much the same way as the drive to the Columbia Hills had over at Spirit's site. Once the primary mission consisting of the first ninety sols was passed, the desire was to expand the access to the sedimentary section in the Meridiani Planum sediments by visiting a deeper crater. Once again, a large orbital mosaic appeared on a table and it showed this lovely crater, informally named

Victoria, eroded into a scalloped-rimmed asterisk shape and with outcrops all around the rim. It was a long way south of the Opportunity site, so it too was thought to be impossibly far away and in some distant future. Looking at the crater in the mosaic on the table, we yet again shook our heads at the impossibility of getting there. A lifetime, or so it had seemed, had passed getting to Endurance Crater just a few hundred meters away from the landing site. How would we ever possibly make it that far? Getting there would be so far in the future that it was just a notional plan. But that too was to come to pass.

We left many tracks or trails in the dust of Mars and it was the type you leave if you are off-roading. Tracks were left in some scenic places occasioning the remark by one engineer that we had better hope the park rangers did not happen along.

But it was no matter. The winds and dust of Mars are very good at erasing our tracks and have done so.

MARCHING INTO THE UNKNOWN

WE MOVED FORWARD LIKE INTREPID EXPLORERS across the new unknown landscape soon after the initial activities at both landing sites. For Spirit, the traverse began once the sol 18 anomaly was behind us. At the time of the anomaly Spirit was in the process of examining the rock named Adirondack, a dark block of basalt a few meters from the initial landing site. It was the first use of the rock abrasion tool (RAT) designed to grind off the exterior weathered surface of rocks, followed by the Microscopic Imager, designed to take up-close images of rocks rather like a geologist's hand lens, and then elemental and mineral analysis instruments. This was all followed up with a look at the results with Pancam, the color stereo-imaging cameras on the rover "head" or mast. We of course all waited nervously on the day that the results of those instruments, including the RAT grind, would be returned. This was the first use of such an instrument on Mars, and so the outcome was a big test of the whole idea that you could prepare a rock surface and do instrument observations from millions of kilometers away. When the results came down, they were better than we'd hoped. There in the image of the rock taken by the Pancam was a big black spot on the surface of the rock with little bits of rock dust blown off to one side. Steve Gorevan, the principal investigator and designer of the RAT, was a happy guy. We were all happy, stunned actually, that the RAT worked so well. The Microscopic

ABOVE: Collection of a few of the tracks left by Spirit and Opportunity in interesting places on Mars over the years.

Imager results showed the first view of a Martian rock up close, so close that individual minerals were visible. The elemental analysis was basaltic, the type of lava that is dark and among the most common on Earth and other planets. We were finally doing field petrology (studying the mineral components and chemistry) on another world.

In the remaining time that Spirit was thought to have—considering that the mission was originally anticipated to last only ninety sols—it was decided to make an effort to visit the rim of a nearby 210-meter-diameter impact crater informally named Bonneville Crater. It was hoped that Bonneville would have sampled something below the basaltic lava covering the Gusev plains and that either the supposed lake sediments that had originally covered the Gusev Crater floor would be exposed in the crater or occur as blocks in the rim of the crater. In this way, perhaps, something of the original plan could be salvaged.

On sol 68 Spirit finally had fought its way to the very rim and the panorama returned at the end of the drive was spectacular. A yawning hole lay before Spirit, and Bonneville Crater proved to be a deep hole in basalt with no rocks from beneath the basalt flows visible. Nonetheless the blocks in the crater rim and the ejecta that were traversed to get there were filled with other details of interest.

The curious thing about a small crater like Bonneville is that all small craters have much the same features, slopes, shapes, and effects on the surface they impact. So a walk on the rim of one is much like a walk on another in terms of the lay of the land. We Earthly dwellers have few opportunities to experience a crater in person, since there are so few well-preserved ones. But I had occasion to relive the experience of being on the rim of Bonneville Crater in a way when I was returning to Albuquerque along Interstate 40 at the end of the mission. After spending so many weeks during the early mission on the slopes of a rather large crater, I thought that I would drop by Meteor Crater—just south of Interstate 40 east of Flagstaff, Arizona—to compare notes with an actual crater. In fact Meteor Crater is perhaps the best example of a relatively fresh impact crater on Earth, so it was a chance to see Bonneville Crater using Meteor Crater as a proxy. Standing on the rim of Barringer Meteor Crater was eerily familiar. The blocky and rocky rim slopes outward to merge with a gentle incline leading off across the crater ejecta, much as it did at Bonneville Crater. Here at Meteor Crater the rocks were sandstones, whereas the rocks on the rim of Bonneville were basaltic. Nonetheless the terrain was identical, with the same drop-off of large blocks with distance outward and the odd sculpted hills and valleys of a chaotic ejecta blanket. I had a sense of just what the situation was at Bonneville as we made our way across the rim, a sort of real-world déjà vu that is difficult to express.

And looking into the crater the view was identical too: a deep bowl covered on the bottom with sediments and rising up all around to steep rocky outcrops just below the rim. It was a sense of complete similarity between two landscapes on two worlds millions of kilometers apart and I was standing in the same setting, one in person and the other by virtue of telepresence. My immediate conclusion standing there in the sunlight was that small craters are very similar wherever you go in the solar system.

Comparison of panoramas taken on the rim of Bonneville Crater on sol 68 by Spirit **(TOP)** and on the rim of Barringer Meteor Crater, Arizona **(BOTTOM)**, by the author. Bonneville Meteor crater is 210 meters in diameter and fourteen meters deep. Barringer Crater is over 1.7 kilometers in diameter and 170 meters deep.

Opportunity's success with exposed layers inside the tiny Eagle Crater suggested that the nearby, 150-meter-diameter and 20-meter-deep crater named Endurance Crater would prove equally fruitful at seeing below the surface of the flat Meridiani plains, so it began a journey there as soon as it climbed from Eagle Crater. After arriving on sol 95 and a brief survey around part of the rim of Endurance Crater, it was decided that Opportunity could climb into the crater and explore an actual geologic section exposed in the crater inner slopes. The epic drive by Opportunity into Endurance Crater was certainly a significant event in the mission. This was the very first time that a rover would drive into a crater on Mars. Would we be able to get out? Was the footing too unstable for doing anything once inside?

There were many decisions and some practice runs with the test bed rover to determine if this was really something that could be done. The slopes were moderately steep, and, as anyone who has tried to walk on a steep outcrop covered in a thin veneer of grit and soil can attest, footing can be slippery. The idea was treated as a mission risk, with all the meetings and discussions that any potentially risky activity entails. Engineers had to sign off on the idea, and tests had to be done. Engineers do not guess, they measure

BELOW: Image of sedimentary layers of Burns Cliff acquired by Opportunity between sols 286–294 while exploring inside Endurance Crater.

and conclude whether something is within the parameters supported by the equipment. It was a process that was repeated for several events throughout the lifetime of both rovers.

The sol came finally—sol 133—when Opportunity did a slight "toe dip" over the lip of the crater rim. This was successful and without an unnecessary slip or other behaviors that would pose a risk of it getting stranded inside the crater, so Opportunity continued gingerly inching down into the crater. Endurance proved to be a success, as it exposed a beautiful section of the layering in the Meridiani Planum sediments forming a cliff, named the Burns Cliff after a prominent planetary scientist who had first proposed that sulfates like those later found everywhere at Meridiani would be common on Mars.

"And so we beat on, boats against the current, borne back ceaselessly into the past."
—F. SCOTT FITZGERALD, *The Great Gatsby*

From images taken at the base of Burns Cliff from sol 286 through 294 it was determined that there were three geologic units, a lower one consisting of ancient sand dunes, a middle unit of thin sand sheets, and an upper unit with characteristics of water-laid sediments. The lower and middle unit were separated by what geologists refer to as an unconformity, that is, the top of the lower unit had been eroded level at some point before the thin sheets of sand were deposited. A similar thing happens in places where there is a shallow groundwater table infiltrating the former dunes that somewhat indurates the sands below the top of the water table and allows the upper drier part to be blown away. Something similar happens at the White Sands National Park in New Mexico, where flat-lying areas between dunes are somewhat moist and a bit crusty because of the shallow water table there. Apparently the water table continued to rise at Meridiani Planum because the middle unit showed evidence of mineralization associated with being relatively soaked in water. When the upper unit was deposited there was surface water capable of moving sand around. All this was determined from a simple stratigraphic section. Opportunity had identified the past presence of significant amounts of water. This was a big discovery.

There were other new world experiences, and these occurred when the rovers crossed into a geologic setting unlike that explored before. Instances when the transition into major new geologic terrains happened for Spirit and Opportunity were transformative moments for the missions. For a geologist stepping across the dividing line—known in geology as a contact—between rocks of vastly different ages, from young rocks and onto very old rocks, it is like a journey through time. In mission parlance we sometimes refer to the arrival at some totally new and different type of geology or different geologic setting as "starting a new mission."

There were two special cases in which one day we were looking at rocks deposited or erupted somewhere in the middle of Martian geologic history, and the next day, after a short drive, we were on a new geological surface that represented the earliest history of Mars. The early history of Mars is where the action is with regard to water and habitable climates. Getting to check it out twice in one mission with two rovers is insane. These were situations in which we learned that, after all those years of looking down from orbit and inferring from the landscapes that ancient times on Mars were wetter, it was true: ancient times were confirmed by the rocks to have been wetter.

Spirit's New World. For Spirit, its second "new mission" experience occurred after we had landed some 156 sols previously. During the previous 156 sols Spirit had driven across a plain of broken lava flows to arrive at the base of the Columbia Hills. But on sol 156 Spirit drove the last few meters from the plains onto the new landscape of the Columbia Hills. We had been driving relentlessly watching the Columbia Hills grow incrementally with each downlink during each sol of that drive. Stepping onto the new, smoother but more hilly landscape after that rocky journey across the plains felt very much like we were arriving on the shores of a new continent after a very long and choppy sea voyage. The contact between the "sea" of lava flows in the plains and the slopes, the "shore," of the new land was like the edge of the surf lapping onto dry land. On the one side were kilometers of sameness all the way to the horizon, and on the other we were standing at the foot of a new landscape rising up to "mountainous peaks." And this was not just a contact between two different geographic areas, this was a contact between two very different geological terrains of two different ages. Those lava flows were old and smashed up by impact craters and three billion years of weathering in the Martian climate. So the Columbia Hills were even older and from a time of the earliest Martian climate, when things—from all appearances elsewhere on the planet—were wild and wet. And the rocks there showed evidence of that ancient wetter time. In one sol of driving, everything about the mission changed, because the Columbia Hills are an old terrain that was already eroded and rounded by the time the lava flows that made up those plains were formed. The long death march across the plains had come to an end and the exploration of a new world began. Driving from the plains on to the slopes of the Columbia Hills also marked the first time that a rover drove across a significant geologic contact on another planet.

Crossing the contact between the lava flows that make up the plains and the old rocks of the Columbia Hills was an unheralded Mars exploration milestone. One of the basic tenets of Mars geologic history from analysis of thousands of orbital images is that there has been a change in climate from early Mars to the Mars of today. The science of planetary geology had been able to make profound assessments of surface geology on other planets just by establishing the sequence of events that have shaped the surfaces we see today. And the sequence is easily determined by overlapping relationships. In this case we were driving across lava flows formed in the middle of Martian geologic history, a time period known by Mars geologists as the Hesperian Eon (about 3.7 billion and 3.0 billion years ago). The rocks of the Columbia Hills are from the time known as the Noachian Eon, a time between 3.7 and 4.1 billion years ago. The Hesperian Eon was a time of many volcanic eruptions, lava flows, and flowing water. The Noachian Eon was a time when there was much erosion of the landscape. Many planetary geologists had reason to suspect that Mars during the Noachian was like Earth 3.8 to 4.1 billion years ago, a time that on Earth includes the era that geologists have called the Archeozoic, which is just about as far back in time as you can go on Earth using the oldest rocks. This is deep time and just the beginning by Earth standards, but a time of great geologic activity on Mars.

Opportunity's New World. At about the time that Spirit was in its last sols of life on Mars, Opportunity was within a kilometer of leaving all the sediments of Meridiani Planum behind and approaching something that I could get excited about. This was about to be Opportunity's second new mission experience. It left the sands of Meridiani Planum and arrived at the rim of the twenty-kilometer-wide Endeavour impact crater on sol 2,680, and on sol 2,681 Opportunity drove from the sulfate sand "sea" of Meridiani Planum onto the "shoreline" rocks of Endeavour Crater, a terrain that was a good billion years or so older than the Meridiani Planum sandstone "sea." In that one drive Opportunity "stepped" back in time a billion years to an earlier, wetter, more dynamic Mars, and it stepped from a terrain dominated by rock debris and sediments, to a terrain that had been dominated by lots of water.

Once Opportunity arrived at the rim of Endeavour Crater I began to refer to the fact that Opportunity had escaped its "sandstone prison" in joking retaliation for Dave Des Marais's comment early in the mission that Spirit was trapped in a "basaltic

ancient Columbia Hills
lower slopes

Gusev Crater floor
lava flows

ABOVE: Part of a Navcam panorama looking south taken by Spirit on sol 156 after arriving on the "shores of its new world" in the Columbia Hills. The lava flows on the right date from the middle age of Martian geologic time, the Hesperian Eon, whereas the rocks of the Columbia Hills on the left date from the earliest geologic time on Mars, the Noachian Eon.

prison." A few team members found that funny—a few did not. The rim of Endeavour Crater was a "new mission" and a new world for Opportunity. The crater rim would prove to be the oldest type of rock that any rover or lander had ever visited on Mars. And "old" in this case means "very earliest time" on Mars. Of course, the very earliest time on Mars is the time that we are interested in simply because that was when Mars was wettest, as far as we can tell from several decades of doing remote sensing of the surface from orbit.

The rim of Endeavour Crater is also a different type of geology than we had ever visited on any planet, even on Earth. Simply put, nobody had ever done ground truth observations on the topographic rim of a crater that size or larger. On Earth we have the small Barringer Meteor Crater and it is the go-to crater on Earth for anything having to do with the surface expression of impact craters. This is simply because it is so young that it is preserved pretty much in the state it existed at the time of formation. Everything we think we know about the details of the rims and impact ejecta on the surface of impact craters is based on Meteor Crater. But small craters are different from big craters. Once the size of a crater formed on any planet exceeds a certain size all sorts of exotic things start to take place when the crater forms, and we had no idea what the geology would look like on a big crater rim. While it is true that we have impact crater structures on Earth the size of Endeavour and bigger, they are all old and severely eroded and lacking the original rims.

The orbital images of the rims of big craters on other planets had shown that they are different from those occurring in small craters like Meteor Crater. The walls of large craters tend to be terraced by giant slumps that form early in the crater formation process or immediately after everything is over but the shouting. Then there is the ejecta of large-impact craters. It behaves more like the welded ash-flow tuffs of large super-volcano eruptions on Earth. So much energy gets released when an object several kilometers across slams into a planet at interplanetary velocities, that the rocks are vaporized, severely melted, and just plain busted to bits all in one big event. And the crust at the scale of whole mountains gets crammed back and peeled up in complex ways. The rain of debris from the explosion is violent, hot, and thick. The resulting deposits and crater rim show all sorts of molten rock behaviors. The final result at the crater rim is some serious faulting, folding, and blast debris, otherwise known as breccia, accumulation. Large crater rims are the province of hard-rock and structural geology. But we have never had a large-crater that we could see at outcrop scale. Here we were with Opportunity after a "sea voyage" across a sand sea, sitting on the shore of this new world, the rim of this unique new type of geology never before seen on Earth. So for many reasons, but especially because of the presence of very old Mars rocks and a very new type of geology, the arrival at the rim of Endeavour Crater was a "new mission" all over again, and we were tiptoeing gently into this new world with eyes wide open for any possibility. In a quiet revolution, any preconceptions about the rims of large craters that we all had were thrown out the window, and every sol was now a new learning experience.

ABOVE: View of Opportunity's new world, the hills forming the rim of Endeavour Crater after arrival on sol 2,681. This is a false-color image with colors stretched to highlight contrasts.

NINE

Adventures on the Red Planet

"Real scientists do not take vacations.
They take field trips."

—EDWARD O. WILSON, *Letters to a Young Scientist*

IN ANY EXCURSION INTO THE NATURAL world there are sublime moments of epiphany, long moments of expectation, and even longer moments and days of slogging in a weary, semiconscious auto mode. But there are times when you are doing the routine of collecting new science, and then in one drive something remarkable presents itself that had not been previously anticipated. These I call epic moments, and we had many of them, some great and some small, but revealing nonetheless when they occurred. When you are exploring a new world every day is an epic moment, but there were some that were particularly grand. There were a few events that revealed either something new about Mars that we had not considered, or if we had considered it, it was too far-fetched to be believed possible on our first mission of exploration on the new world. And then there were some epic moments that revealed a new thing about operating a rover on another planet.

EPIC EXPEDITION MOMENTS

THE MIRACLE OF SOL 419

A REMARKABLE THING HAPPENED TO SPIRIT shortly after it had survived its first winter on the lower slopes of the Columbia Hills. It had begun the process of climbing up the slopes, but power was now an issue as the dust had accumulated on the solar panels over the previous four hundred sols to such an extent that the available power was seriously lower than early in the mission. The dreaded reduction in power from dust that we thought would kill us was creeping up on us. But the other power issue was the low Sun of winter, which on top of the dust problem should have been the final showstopper. However, we had dodged that bullet rather cleverly during the winter by perching Spirit at an angle on the slopes during the middle of the first winter so that the solar panels faced the Sun. It was a thing that had not been seriously considered before the mission because who could predict that we would have slopes of the magnitude necessary to tilt toward the Sun? But we were on a mountain, and mountains have slopes. This worked very well, and we used that repeatedly during the rest of the mission. We continued to park Spirit on a slope every winter with magnificent results. In doing that we had learned one of the first lessons about surviving on Mars with solar power. More important, we learned that being powered by solar panels was not the terminal illness that we had spoken about early in the mission. We were beginning the painful process that every pioneer must learn, how to live off the land and make use of its characteristics to your advantage.

We were beginning the painful process that every pioneer must learn, how to live off the land and make use of its characteristics to your advantage.

On this occasion we were finally approaching the outcrop on Larry's Lookout on the slopes of the Columbia Hills; we had stopped on sol 418 to do a survey of the outcrop before approaching it in a lead-up to doing a short campaign of observations on the rocks. The next day we would get the downlink with the necessary documentation of the scene and would follow that up with an approach to the outcrop, where we would position Spirit for touching the rock face. The next morning, during the SOWG process in which there is a report on the health of the spacecraft—including power and other factors that affect the planning—we had a little surprise. On this particular morning, the engineering team revealed that the power output from the solar panels was back to the levels at or above those we had on the day we landed! Somehow the solar panels had been cleaned of their unwanted decoration of Martian dust.

But how? A quick look at the available images gave a clue. The front Hazcams, the lower-resolution cameras that looked out from the front of the rover directly in line with the instrument arm, were cloudy with dust smears. It was as though something had kicked up all the local dust and blew it into the rover's "face." Eventually to check what was going on we commanded a survey of the solar panels and the deck looking down from the panoramic camera mast, and there we saw, in addition to shiny brand-new panels, swirls and streamers of dust hither and thither across the superstructure.

Clearly a big wind had come along and swept the panels free of dust. Thus we had learned that what the Martian atmosphere giveth, it also taketh away, if you are lucky. Natural solar-panel-cleaning events obviously happen and in the process breathe new life into a mission that might otherwise head downhill fast with excessive dust accumulation.

Spirit sol 330

Spirit sol 432

ABOVE: Comparison of "selfies" taken by Spirit of its dust-covered solar panels on sol 330 and again on sol 432 after the panel-cleaning events of sol 419.

This was to be the first of many solar-panel-cleaning events throughout both missions, and in this way we were given an extended lifetime far beyond the predictions, and for free. All you had to do was wait and, out of the blue—or rather on Mars, out of the mauve, given the dusty color of Mars's atmosphere—you would be given new life. Panel-cleaning events were not entirely predictable, but we were to learn over the years that in the warmer sols of spring and summer regular gusts of wind were common. Winters were the worst, since the atmosphere was quiet and still, and over the course of a winter the dust situation would frequently begin looking grim. But every spring and summer, the winds would do the job.

The magnitude of these panel-cleaning events later became a standard part of the long-term planning leads, daily report in a serial plot of what was known as the dust factor. That was simply a visual device for showing trends in the amount of dust on the panels. A high dust factor was good. Generally it trended gradually downward as dust accumulated over a sol-by-sol basis, and then one sol you would wake up and the factor was high because the wind had once again done its job.

Whenever I give a presentation to the public about the missions, one of the first questions that comes up is why we did not put some type of panel sweepers on the panels if dust was such a problem. Of course cost, weight, complexity, and concern for more deadly matters than lack of power were the main reasons for not designing that into the spacecraft. But I usually concluded by simply pointing out that we do not need a mechanical panel cleaner, the wind does a fine job of cleaning the panels and it's even free, if not entirely reliable.

SHADOW SELFIES

LONG BEFORE SMARTPHONE SELFIES BECAME A thing, both rovers had a habit of capturing selfies of the shadow kind during their respective traverses. There were many examples of this simple lighting effect. But they all elicited the same awe that here we were actually seeing evidence that these scenes were being taken by real Earthly hardware in adventurous unearthly settings.

Spirit got quite good at it early in its march across the plains from the landing site to the Columbia Hills. The drives were in a general heading of east-southeast, and late in the sol, at the end of each drive, a Navcam panorama was acquired for purposes of planning the next sol's activities and drive. Because the Sun was low in the west-northwestern sky the shadow of the rover was frequently cast toward the southeast in the general direction of the next drive, often with an angelic halo of antisolar backscatter surrounding the camera shadow on the mast. There was something anthropomorphic in the way each day the rover shadow was cast on an interesting scene nearby as if to say, "Look, here I am in this amazing place."

Then there was the epic shadow cast by Opportunity as it worked its way into Endurance Crater. The front Hazcams, the cameras mounted low on the front of the rover, were also part of the standard end-of-drive images acquired just in case there was something in front of the rover that might prove either hazardous or interesting for further analysis. In one Hazcam image the interior of Endurance Crater was clearly filling the scene in the soft light of afternoon, and splashed across the scene

BELOW: Spirit views its shadow on the long road to the Columbia Hills seen in the far distance in this sol 124 post-drive image. Spirit would go on to eventually summit the highest peak in the Columbia Hills.

Opportunity sol 180 Endurance Crater

Opportunity sol 3051 Endeavour Crater

Opportunity sol 2381 Meridiani Planum

Opportunity sol 3020 Endeavour Crater

LEFT: Opportunity was an experienced taker of "shadow selfies." Over the course of the mission many such images were acquired as panoramas, which were taken at the end of each day's drive. **ABOVE:** Long shadows cast across the landscape at the end of a field trip to a Mars-like landscape in New Mexico are like the shadow selfies that Opportunity and Spirit saw at the end of their drives on Mars.

was an enormous elongated shadow of the rover camera mast and wheels. The effect was almost as good as a picture taken of the rover sitting in this alien scene with the words scrawled across it saying, "Wish you were here."

Often, I have taken similar shadow selfies when working in the field in New Mexico when the late-afternoon sunlight imparts that Martian character to the landscape and my long shadow projects out around the terrain. Whereas many people might find the landscape alien, these types of scenes on Mars would seem to a traveler from the American Southwest imminently reminiscent of something seen at the end of a long day in the field just as the Sun was about to set. It is all part of the abiding sense that experiences on Mars remind us of experiences many of us have had here on our own planet. That long-shadow selfie, whether intentional or not, can be a satisfying reminder of a day well spent in a beautiful landscape.

THE FIRST MARTIAN SUMMITING

SPIRIT WAS THE FIRST ROVER TO summit a "mountain" on another planet. Officially the goal was to explore the Columbia Hills from bottom to top, and that meant getting to the top. Layering, so much sought for in a mission focused on the rock record, was certainly encountered during the climb. But instead of horizontal layers stacked like the pages of a book, these layers were almost conformal to the slopes. It was as though layers of material had been draped over the hills over the course of Martian geologic history. Different events could do this, including ash from the many volcanoes in early Martian geologic history, or the ejecta from nearby impact cratering events. Whatever the origin, it was clear from chemical analysis that many of these materials were severely altered chemically by water in the past, either before being emplaced or over the course of subsequent years when snow or groundwater repeatedly soaked the deposits. Whatever the origin of the deposits it was also clear that they were steeper than the slopes, as though the layers on the upper part of the mountain were stripped off over time so that outcrop layers lower on the slopes were tilted more steeply than the actual terrain slopes and exposed in outcrop edge on.

dust devil ←——

Spirit wended its way through this interesting landscape identifying a variety of different layers, and all were tilted the other way as it descended the other side of the mountain. There is a rule of thumb in geology: when you have a stack of layered materials conforming to a hill shape, then erosion of the top of the hill exposes deeper layers. In order to see the maximum geologic section possible, it was necessary to summit the mountain. On sol 581 we had finally scrambled to the nearly level terrain within striking distance of the summit. As Spirit arrived just below the summit it took the standard end-of-drive Navcam panorama, and the view was epic. There to the north was the long ridge of the Columbia Hills surrounded by the plains clear out to the north rim mountains of Gusev Crater. More epic still, there was a large dust devil twirling past the Columbia Hills. The only thing missing was the wind flapping our gear as we stood near the summit.

But the true summit lay a few meters toward the east and before we could visit the summit it was necessary to inspect the rocks in an outcrop just below the summit that form a slight platform. The rocks were largely horizontal instead of tilted like the outcrops on the slopes, although the particular part of the outcrop that we visited had long ago broken and cantilevered at an angle. Since we were using a mountain-climbing naming theme, this rock was named Hillary, after Sir Edmund Hillary, while Tenzing Norgay was the rock at the very summit. The rocks were similar to some of the outcrops that we had seen lower on the slopes. So instead of deeper layers being exposed at the summit, it was apparent that these outcrops, being laid down

LEFT: Opportunity's sol 3,894 "summit panorama" from the highest point in its traverse, the summit of Endeavour Crater's rim, "Cape Tribulation." The crater floor stretches twenty-two kilometers to the far rim in the distance. The rover's arm was held up into view (INSET) carrying the United States flag on a small metal motor shield made from a piece of the World Trade Center.

somewhat horizontally on the summit, had acted like a cap and were somewhat protected from the erosion seen on the slopes leading up to the summit. Thus it appeared true that the surface of the Columbia Hills had been draped with later materials and these rocks were draped on the more or less flat-lying summit.

There was a slight step just below the summit somewhat reminiscent of the famous step below the summit of Everest, but once the engineering team and rover drivers ascended that, it was then an easy matter to back up onto the true summit. Hence Spirit became the first rover on another planet to ascend to the summit of a "mountain." The view was awesome. A 360-degree panorama was of course necessary from this ultimate perch, and all around lay spread out the floor of Gusev Crater, the plains that we had driven across to arrive here, and far away the rim of Gusev Crater. And to the south in the valley lay the enigmatic Home Plate structure that was our ultimate target. And above that extended the southern ridges of the Columbia Hills. All that was missing here was planting a flag. The engineering team was happy and rightly so. It was another interplanetary rover first.

Then someone noticed an odd thing about that peak to the south. All around us we could see the horizon far away and you could put a straightedge on it. This straight horizon marked the zero-elevation angle as seen from Spirit. But the next

peak to the south was sticking up above this horizon. What? Wait a minute, if that peak is sticking up above the horizon, then doesn't that mean it's, well, kind of higher than the peak we are on?

Using some simple trigonometry, measuring the elevation angle of that summit using the Navcam, and the known distance to that summit from the orbital map, the southern peak was estimated to be a good sixty meters higher! This, then, was an interplanetary face-palm moment. We had not climbed to the true summit of the Columbia Hills; the next peak to the south was even higher. In the words of a famous cartoon character, "Doh!" But we had climbed to the peak of this mountain, and that was good enough for now. In this way Spirit was engaged in some planetary survey work, determining things about the relative elevation of peaks for instance that are difficult to do unless you are actually on the ground.

Opportunity did its version of a summiting during its drive along the rim of Endeavour Crater. At one point along the rim it ascended to the highest segment of the crater rim, which was named Cape Tribulation, and that was the highest elevation along the entire forty-five-kilometer traverse. Again, a panorama was in order for this epic moment looking over the crater floor toward the far rim some twenty-two kilometers away. To make the moment complete, the rover instrument arm bearing the rock abrasion tool (RAT) was stuck up in front of the camera to display the small American flag affixed to a plate on the outside of the RAT. You always need to plant a flag when you first ascend a summit.

A little-known side story is relevant about the plate on which that flag is affixed. Honeybee Robotics, the engineering company that built and operated the RAT, was located in New York and completing final assembly of the device when the events of September 11, 2001, unfolded, destroying the World Trade Center just a few blocks from the Honeybee Robotics offices. Steve Gorevan, the RAT principal investigator from Honeybee Robotics, was able to acquire a small piece of metal from the debris of the World Trade Center that he then fashioned into a small plate to shield the RAT motor from dust and debris. The plate with the American flag in the summit image, then, is actually a small piece of the World Trade Center that went to Mars.

A DREAM COME TRUE AT HOME PLATE

AS SOMEONE INTERESTED IN THE VOLCANOES of Mars, I have always lamented the fact that the search for evidence of past habitable environments in the rocks always takes landers and rovers to nonvolcanic areas, unless by accident. But lava flows like those on the floor of Gusev Crater are one thing and actual volcanoes are another. Visiting a volcanic vent was an impossible dream that might happen someday, perhaps in the distant future when we might actually send a rover or even a human mission to a volcanic area. By what chance could the first mission to drive across Mars—accomplishing many firsts along the way—traveling in what was not particularly known as a volcanic area of the planet, hope to be the first mission to see an actual volcanic vent (the actual edifice of a volcano) on the ground? It would have had to have been outrageously fortunate. However, nestled in an obscure valley of the Columbia Hills, was something that looked like a volcanic vent structure. But was it a true volcanic vent rooted in an eruption of magma from the mantle, or was it something else?

Home Plate was a destination from the early days of Spirit's mission because it appeared to be an anomalous white irregularly shaped disk some ninety meters across nested in the bottom of a valley. Many on the team initially interpreted it as a possible evaporite deposit like one might expect in a dry lake or pond. Given the mission's focus on past water environments, it became a high-priority mission objective. Finally, after climbing up and over the Columbia Hills and descending into the valley where Home Plate lay, Spirit was on final approach around sol 744.

ABOVE: MRO/HiRISE image of Spirit's traverse around Home Plate and an overhead view of Yankee Stadium at the same scale.

Home Plate appeared to be a low mesa-like feature with outcrops around its margin. In an image taken on approach, one particular outcrop on the northeastern edge appeared to have thin horizontal bands or layers, and excitement mounted. Jim Rice, one of the science team members, was of the habit of scanning daily images after a downlink and pointing out all manner of interesting things of possible importance, often to the consternation of those with plans for some particular target and really not wanting any wrenches thrown into the plans for getting to those targets. But on this day, Jim posted a picture of the outcrop in question and suggested that there was a "bomb sag" in the layers. Now, a bomb sag is something that occurs in layered volcanic-ash deposits, often around an explosive volcanic center, and the term "bomb sag" comes from the fact that frequently a large volcanic ejecta block, referred to as a volcanic bomb, actually makes a dent or sag in a deposit when it lands forcibly after its trajectory through the air. Bomb sags are often revealed when later erosion exposes an outcrop of volcanic ash or pyroclastic deposits in sections.

Classic "bomb sag" type structure in the ash layers on the northwest margin of Home Plate **(BOTTOM)** compared with a similar feature in the ash layers of Zuni Salt Lake volcanic crater in western New Mexico **(TOP)**.

The surrounding area consisted of many volcanic rocks densely pocked with gas bubbles (bubbles that geologists call vesicles) as one would expect at a place where there were many volatiles and ash deposits, including some deposits that consist of small spheres that occur when molten rock gets sprayed into the air.

To make matters more confusing, the layers in the side of the Home Plate "mesa" consisted of two parts, a lower part of volcanic ash and the bomb sag and an upper part that analysis showed was simply the same bedded ancient sand dune materials that we had seen elsewhere. The debate about whether Home Plate was a volcanic vent or just an old crater that had been filled with ash from some distal place and overlain with dunes raged within the science team. This is the classic situation of the "argument on the outcrop" familiar to all geologists who have gone on a field trip to some exposure that held ambiguous clues. But at ninety meters across, to many of us Home Plate was probably too small to be a volcanic vent formed by eruption of magma from deep within Mars for many reasons having to do with how volcanoes work. So what was it?

Some of us eventually speculated that it might be a type of volcanic-explosion crater that occurs when hot lava flows over water-saturated ground. In fact, there are other somewhat circular structures in the vicinity of a similar size, and two very intriguing features just to its south that appeared to be more completely exposed examples. The lava-water interaction explosions of that type on Earth also tend to occur in clusters so that much was consistent with the idea. And in this case the cluster occurs right at the edge of a set of earlier lava flows that had covered the plains and lapped up onto the slopes of the Columbia Hills in a low alcove. But the evidence for the whole idea of lava-water interaction was never sufficiently tested before Spirit's mission came to an end. It is unfortunate because in the end that would have meant that Spirit actually visited a place that had lots of water when the lava flows occurred, and evidence for the past presence of lots of water was of course one of the holy of holies for the mission overall.

So in the end Home Plate turned out not to be an evaporite deposit but was likely something even more spectacular in terms of the quests for ancient environments of water on Mars. Another twist by Coyote Mars.

THE STRANGE CASE OF DARK SAND PUDDLES AND FLOWS

IT WAS DURING OPPORTUNITY'S EXPLORATION OF Marathon Valley's south wall that we were examining the local rocks on a steep slope and noticed that there were odd dark sand puddles on the ground. Marathon Valley was so-named because it was at this point in Opportunity's traverse along the rim of Endeavour Crater that it exceeded a marathon distance of travel on Mars, 42,195 kilometers. The puddles were a few centimeters wide and looked for all the world like they were actually emerging, perhaps "erupting" is a better word, from the ground and flowing a few centimeters downslope. These were fresh appearing because they were not coated with the ubiquitous dust that coats anything older than the day before on Mars. Was the sand somehow flowing out of cracks in the ground? Why did they only appear when we showed up with Opportunity and started poking around? It was an unsolved mystery even after a brief campaign using the Microscopic Imager to look at them in detail. The puddles were tongues of dark sand, and the sand itself was similar to that occurring throughout the landscape. So we moved on in the collective need to keep moving, scratching our heads at yet another Martian oddity to be mulled over in the future using what little analysis we could do with our instruments.

A bit later Opportunity was attempting to get up close to the rocks on the valley wall when it exceeded 30 degrees of slope. Of course, 30 degrees is close to the angle of repose of loose materials and so at that point Opportunity simply started digging in without further progress. In fact, it was the steepest slope that Opportunity had ever attempted, and progress was being made until the last bit when the wheels started spinning. The outcrop it seemed was going to be just out of reach. So we grabbed some images at close quarters and prepared to back away. The rover drivers, ever cautious, wanted to make certain that there was no obstruction or other hazard to backing up at this point, so we acquired a few images with the cameras on the rover mast looking backward over the deck. It was there that we saw dark streams of sand that had flowed down off the solar panel. It was then that science team member Rob Sullivan suggested an explanation for some things. Rob followed the tenets of good science by often citing multiple working hypotheses during weekly science sessions whenever we were having a go at some vexing question, and on this occasion he

LEFT: Streams of sand draining off Opportunity's solar panels on sol 4,322 after a climb on a 30-degree slope. **RIGHT:** An image from sol 4,262 of one of the unusual dark sand puddles sitting on the surface. The sand appears to magically start at the top (arrow) and flow downslope.

pointed out that all that dark sand flowing off the deck while we were on the slope might account for the odd puddles of flowing sand that we had recently seen. Clearly the sand on the deck was exceeding its angle of repose, even if the rover was not, and streaming onto the ground in one spot like an open-air hourglass. It was the only explanation that fit all the puzzling characteristics. In fact Jeff Johnson, the Pancam team camera spectroscopy guru and a tall, calm, even-toned, and businesslike sort, had been in the habit of showing us many instances previously of "some strange dark substance" that moved around between the solar panels on the rover deck between subsequent pictures. He even made a movie of it, which was eerie because the stuff was absolutely black and moved around in the movie like some alien amoeba crawling over the rover deck. It was the dark sand that had collected on the deck and moved from place to place as the rover bumped along its roadless traverses.

But how did the sand on the decks get there in the first place? Ron Greeley from Arizona State University—a pillar of the planetary sciences from the beginning, present during the Viking missions way back when, a fellow traveler on the road to Mars over the years, and supporter of many graduate students who went on to become eminent planetary scientists in their own right—had given us a clue a few weeks earlier. Ron was an expert in windblown features and their origin on Mars and was observing some of the small ripples and dust devils throughout the mission. He had given a little presentation in which he noted the sand on the rover decks and discussed how it was evidence that the process of sand movement on Mars was a bit livelier than we had come to expect. His interpretation was that during wind events the grains of sand were actually bouncing as high as the rover deck and some of the sand was collecting on the rover. In the early days we had thought that sand was unlikely to be mobile on Mars because the force necessary to move something larger than dust would be much greater than the thin atmosphere of Mars could provide. In fact, it was assumed that all dunes and ripples would be made of dust.

But our early poking at some ripples over on Spirit had shockingly shown that the wind ripples were made of sand, so it must be possible after all. Ron had advocated the idea that the movement of sand was facilitated by a process known as saltation in which a short hop by a grain bounces another grain into the slip stream and so on and so on. In the end sand could be moved quite easily even in the thin Martian air. The whole thing was obviously a little sleight of hand on Coyote Mars's part just to keep us off-balance with thinking our fancy physics worked in simple ways. The strange case of the sand puddles and flows was solved through chance when Opportunity tried to climb a steep slope.

Opportunity ultimately exited the valley crossing over into the next valley and eventually wended its way back to the rim after a steep climb up an adjoining valley. From there Opportunity embarked on a short journey over the next rim crest to the south, where it approached the next big mystery: Perseverance Valley, to be discussed in a later chapter.

THE STORY IN THE ROCKS

MANY OF THE DISCOVERIES ABOUT THE rocks of Mars were not epic in the sense of being significant events in the grand narrative of expeditionary exploration, but there were many instances in which surprising results occurred from a simple rock or outcrop. This was a mission to learn all we could about the past climate and environments of Mars, and any field geologist will tell you that the best way to do that is to just look at the rocks. And if you can look at them in their original outcrops all the better because the context of a rock is like a biography. The minerals, their grain sizes, and the structure of a rock all record epic stories about how the rock formed. Stories can be fiction, or if they are based on facts, the stories are called histories. Rocks are good at recording histories of long-ago times. So we spent much time seemingly obsessed with rocks in general and outcrops in particular.

In past landed missions we had looked at the rocks in the immediate vicinity of landing, but for the most part the rocks were out of context and just lying on the surface. One could make some inferences based on the region where the rock occurred and what we could infer about the geologic history of that region. Previous landing sites told us that many of the rocks came from somewhere else, usually upstream on a major outflow channel. This meant that there was insufficient information to determine the exact geologic history of the rocks or where those rocks were formed in the original timeline of Mars geologic events. But seeing a rock in the setting in which it occurs, an outcrop, provides a short biography of the rock, including when it formed in relation to other rocks and, in many cases, what has happened to the rock since that time.

A single isolated rock is like seeing a person sitting in a chair in a featureless room. You can make some inferences based on the characteristics of the individual,

> This was a mission to learn all we could about the past climate and environments of Mars, and any field geologist will tell you that the best way to do that is to just look at the rocks.

but they are generalized and you would not know anything about who the person is, what the person was really like, or anything about their history and how they got to be who they are. But if you saw the person in their daily setting or maybe a short video of them engaged in something that they do, then you start to learn something. Or if you are Sherlock Holmes you can use clues that you have learned to infer from a single footprint whether the person was five feet or six feet tall, or maybe whether the footprint was made by someone in a frock coat or carrying a chimney sweep's broom. It seems impossible, and some of Holmes's deductions were, but with enough experience and knowledge of how things work, you can determine some details that appear from outside as impossible. Rocks are the same. Each is an individual, and each has a story to tell. But since rocks cannot speak, it is necessary to do some biographical research from the clues. This means touching them, "tasting" them with chemical analysis instruments, and looking at them up close with a Microscopic Imager.

DESERT VARNISH AT MAZATZAL AND ROUTE 66

AROUND SOL 77 ON SPIRIT, WE were moving along the south rim of Bonneville Crater and concluded that since we had not examined a rock for some twenty sols and in the interest of sampling the potential diversity, it was time to analyze another rock. We identified a rather large flat boulder sitting on the rim of the crater that appeared to be well oriented for examination. Since we had been naming large rocks after famous mountains, we named this rock Mazatzal after a mountain near Phoenix, Arizona, that is somewhat well known in geological circles as a type area for ancient rocks in the Southwest. On this rock we would deploy the full suite of activities, including initial analysis with the chemical tools, followed by brushing and then grinding with the RAT to get a clean surface free of alteration and dust. These were always followed up with a look at the results using the up-close Microscopic Imager.

As before at the rock Adirondack, the first brushing exposed what appeared to be a very dark surface, almost black. But after grinding, as before, a somewhat brighter dense basaltic interior was exposed. Except in this case the surface was not quite flat and only half the grind circle was actually ground down to clean rock. So we applied the RAT grinder again, and the results were rather interesting. A thin strip across the face of the grind had still not been ground, but that led to something curious when we looked at it with the Microscopic Imager: it was clear that the unground black strip left behind was a coating on the rock.

It was becoming obvious that Mars rocks have a kind of desert varnish beneath their ever-present dusting of light brownish-red dust. The best that we could tell from the chemical analysis comparing the pre-grind and post-grind measurements was that the varnish was mostly sulfates. Curious, but we needed more information. One example is not sufficient to come to a conclusion.

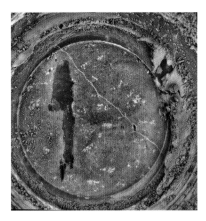

ABOVE: Microscopic Imager documentation of a RAT grind on the rock Mazatzal on sol 82. Color was added from the Pancam image on sol 85. The grind circle is four centimeters across. The dark vertical band is an unground part of the original rock surface retaining the pre-grind "varnish."

Perhaps a little more information occurred later around sol 99 as Spirit began its great roadless road trip to get to the Columbia Hills. I recall sitting on one of the many press media panels or press conferences that the mission held early on. At this particular press conference I explained that while Spirit was about to do a long drive across the plains, this was not going to be some sort of long, dull, bad road trip, but instead we would be stopping along the way periodically to see the sights. Just after descending from the outer slopes of Bonneville Crater on the start of that road trip, we stopped to investigate a nice flat-faced rock that we named Route 66. Obviously, the name was chosen because we were using a "road trip" theme at that point. We had a few sols to spend at this location so it was decided that we would do a little experiment looking at the surface with the Mini-TES instrument, the device that used spectral characteristics of reflected light in the near-infrared to determine things about mineral compositions of the surface after brushing the rock. The problem is that the area that the RAT brush would clean with one grind was only about four centimeters across and the area viewed by the Mini-TES, its observation "footprint," was considerably larger than that. As a result, the Mini-TES instrument would not be able to get a clear signal of the surface without the dust.

It was at this point that Steve Ruff had an idea. Steve was the deputy principal investigator for the Mini-TES instrument and was always on the lookout for ways that it could be used to add to the science goals. Steve is an intense guy who frequently barged into the uplink planning sessions with a comment like, "Wait a minute, we are going about something wrong." He often acted as a science conscience for the team by pointing out why something should be done and that if we didn't do it, we were

TOP: The rock Route 66 acquired on Spirit's sol 100 shows just how dusty things are on Mars. The unusual pattern is the result of a series of brushing actions by the RAT to remove dust. It exposed the characteristic shiny "varnish" that occurs on many Martian rocks.

ABOVE: Petroglyph on basalt in Petroglyph National Monument, New Mexico. Petroglyphs are created by chipping off the dark desert varnish that forms on the surfaces of rocks in the desert Southwest and exposing the grayer rock underneath.

derelict in our duties. Most of us might come in with that attitude and get beaten down with quick-witted arguments to the contrary, usually something about not having the time to spend doing more detailed work on something just because it was an interesting little "science fair project." But not Steve; once he latched on to a strong belief he often hammered on until we relented as a team. The problem was, he usually had good arguments, which was generally somewhat irritating. This even happened to the tune of driving the rover back to a previous site when he felt that we had failed to complete the work there.

On this occasion Steve suggested something that was a clever interaction of instruments. He proposed that we do multiple overlapping RAT brushings of the surface at the rock Route 66, thereby achieving sufficient area for the Mini-TES to "see." As it happened, this was about the time that the Olympic Games were on the minds of the sports-focused folks. Steve proposed as a little addition to do five RAT brushes on the rock in a kind of Olympic logo of three rings in a row with an additional two rings below, thus garnering a little "Olympic symbol on Mars" publicity while also satisfying the need of a sufficient area for a decent Mini-TES observation.

The brushing was done, and the results were rather interesting. With the much larger area exposed it was clear that the dark surface was glossy, even reflective. So, what did the Mini-TES reveal? The results were so-so because the signal was noisy. The spectral signal of CO_2 dominated and nearly washed out the other signals. Apparently, the reflective surface was giving us good spectra of the reflected Martian sky, which of course is mostly CO_2! But there was a sulfate signal in there, and the ultimate conclusion was that the dark coatings are sulfates.

We went on to see these coatings on many rocks, and the idea of a desert varnish-like coating on Martian rocks was intriguing. When we did RAT grinds through the coatings we were effectively doing these circular petroglyphs on Mars rocks. Of course, desert varnish is common on rocks on Earth in desert settings, where actual petroglyphs pecked into the dark coatings were done by Native Americans. I happened to be familiar with the phenomena, living right next to Petroglyph National Monument.

Every morning, I look out at dark basaltic rocks that, while being dark gray on broken surfaces, are coated with a darker brownish-black glossy varnish. In fact, on many an afternoon, as the Sun is high in the sky, the rocks look blue because I am obviously seeing Earth's blue sky reflected in the varnish, much like the Martian sky was reflected in the rock Route 66. Earthly desert varnish is mostly a coating of silica-rich minerals and pretty much the minerals that occur in typical dust that coats everything in the Southwest. The dust, of course, is from the silica-rich rocks, often clay minerals that are eroded from typical rocks that make up the surface.

While the origin of desert varnish is still a matter of research, and some think that a microbial component is involved in its fixation on rock surfaces, I prefer the simple explanation that the varnish is actually a kind of water-based power coating. The concept is that dust settles on a rock and a little moisture or rain fixes it to the surface, rather like crusty dust spots on a car's paint after a misting short rain. Over thousands of years the powder coating builds up into a thin layer of dirty varnish.

But what has this got to do with Mars? One of the speculations was that Mars rocks are perpetually coated with dust, which itself is rich in sulfates as sulfates are everywhere on the surface of Mars. Something, perhaps an occasional amount of water from snow or ice, melts on the sulfate-rich dust and fixes it to the surface. But whereas water or snow is not present today, there is the possibility that every few thousand years a dusting of snow or frost is possible, when the climate is right. Theoretically that could happen when Mars's poles are tilted somewhat from its current orientation. And that can happen because the tilt of Mars is not stable but wobbles, essentially because Mars does not have the stabilizing effect that Earth's Moon provides on our tilt. Over the course of hundreds of millions of years or a billion years, enough of these events could add up to the same effect that we see on desert varnish on Earth after just a few thousand years. It is an unproven theory, but fits with what we currently know about Mars climate variability. So once again we may have found evidence for water, and in this case the water is more recent than we were expecting.

SOFT HARD ROCKS

ONE OF THE INTERESTING DISCOVERIES MADE possible by the use of the rock abrasion tool was the fact that many of the rocks encountered are much softer than their Earth counterparts. By monitoring the motor currents on the RAT grinder, it was possible to establish the hardness in terms of the resistance to grinding. Envision using a handheld sander on a soft material versus something firmer; the motor slows down, and more current is required to keep sanding. The RAT team referred to the force required to grind a certain distance into a rock as the specific grind energy, and recorded a range of values during the course of many rock grinds throughout the traverses of both rovers. By measuring this force on rocks of comparable type, such as basalt or sandstone, in the lab we could see how the rocks on Mars compared in terms of hardness.

> The rocks were perhaps ghosts of their former selves left behind after a long, early history of water soaking.

It was noted with the first grinds at the basaltic rock Adirondack and other rocks on the plains of Gusev Crater that Mars basalts were a lot softer than a typical Earth basalt. Basalt is basalt regardless of planet, and basalts are made of certain minerals and silica glass that are the same everywhere, so the basalts on Mars should be comparable in hardness. One suggested explanation for the apparent difference was the fact that the grinds were only a couple of millimeters deep and that we may have been measuring some type of weathered and soft exterior rather than the interior rock itself. Nonetheless all the basalts visited by Spirit at Gusev Crater yielded low hardness numbers, although it should be added they were still hard enough to wear out the grinding bit by sol 420 such that Spirit was thereafter reduced to just brushing rocks. But by that time, we had a rough calibration of the difference between unground and ground rocks so the possible exterior rinds could be subtracted from results obtained on unground rocks.

Over at Opportunity, in Meridiani Planum, where the rocks were sedimentary and expected to be soft, they were even softer and had all the hardness of dried mud. This was not terribly surprising given that most of the rocks there were cemented by soft evaporite minerals at best. But when Spirit arrived at the Columbia Hills, where the rocks were much older, the rocks were very soft, much softer than the basalts of the plains. The same thing occurred when Opportunity arrived at the rim of Endeavour Crater. These were old rocks again, but were expected to be large chunks of various hard rocks formed during crater formation. But these too were fairly soft.

While it was possible that all the rocks we visited were simply made of soft materials like ash and various low-density broken rock bits, one speculation somewhat supported by chemical analysis was the possibility that the rocks were simply corroded and altered. The rocks were perhaps ghosts of their former selves left behind after a long, early history of water soaking. A similar thing happens to Earth rocks that start out hard, like granite. In the southeastern United States, for example, extensive red clay and crumbly granitic outcrops, a type of rock known as saprolite, dominate the landscape. In this case these rocks started as normal granitic rocks but were corroded over geologic time when they existed in former climates that were tropical. The same phenomena leaches rocks of the more easily dissolved minerals like iron, which rusts the rocks red and turns the more abundant minerals like feldspar, which is common in granite, to clay. In extreme cases the resulting rock is turned to bauxite, an alumina-rich rock suitable for mining for raw aluminum, when the minerals are so degraded that only the alumina that is a major component of feldspar is left behind. Hence the soft "granites" of the Southeast. And perhaps the older rocks of Mars will turn out to be a kind of ghost rock or saprolite. At least the measurements made by both rovers suggest that could be a possibility.

EVIDENCE OF WATER, WATER EVERYWHERE

AS NOTED, A PECULIAR THING HAPPENS to rocks that are soaked in normal, that is, nonacidic, water for long periods of time. The rocks can turn to mush, or at least the rock equivalent, which is clay. Certain minerals that are common in most rocks, and feldspar is an important one, get altered to a variety of mineral known as phyllosilicates. Mica is a well-known example of this type of clay mineral, but there are other varieties in addition to the glittery flakes of mica that form by altering harder minerals. These include more complex compositions and dust-size particles. In fact, a lot of dust is the result of rock weathering and consists of varieties of phyllosilicates. But the one thing that characterizes all the phyllosilicates is the fact that they are the result of water interacting with preexisting minerals, either in groundwater, surface water, or especially in interesting places like hot springs. The presence of clay, then, is evidence that water was abundant either when the rock formed or during the

ABOVE: Image taken by Opportunity on sol 3,230 of an outcrop with a "boxwork" pattern of alteration along ancient fractures in the bedrock on the rim of Endeavour Crater. The image is about seventy centimeters across.

course of alteration by lots of water later. The saprolite I mentioned is a good example of that process. In any case, searching for clays was another way of pursuing the general theme of Mars exploration that was summarized in the "follow the water" mantra of NASA.

It is no surprise that when orbital remote sensing started identifying phyllosilicates, or clay minerals, on Mars, the game was afoot. The Mars Express instrument OMEGA (Observatoire pour la Minéralogie, l'Eau, les Glaces et l'Activité) and Mars Reconnaissance Orbiter instrument CRISM (Compact Reconnaissance Imaging Spectrometer for Mars) are especially designed for detecting different minerals and phyllosilicates are important ones.

For many years it was suspected that if phyllosilicates were going to be present on Mars it would be in the oldest rocks because that earlier time on Mars is the time when all the geological evidence from previous photogeologic mapping had shown that water was present in abundance. And the early detections were showing that to be the case. Clays were being detected in older terrains. That is why one of the divisions of Martian geologic time was named the Phyllosian after the phyllosilicates, or the clay minerals that defined that era. But what were the clays like, and where did they occur? What did the rocks themselves have to say about why the clays were present? Were they former sites of mud, or just highly altered rocks like saprolite? We really needed to look at the rocks in a place where phyllosilicates occurred.

This is where Opportunity comes into the scene. Opportunity arrived at the rim of Endeavour Crater, which was known to be a terrain that formed long ago when Mars was potentially wet, and lo and behold, phyllosilicates were detected by CRISM in a small area on the east side of the rim segment. Once Opportunity made its way near the location where the remote sensing said that there were gobs of phyllosilicates, we drove Opportunity upslope and did some prospecting.

Arriving at the first outcrops, everything appeared in order and it was not obvious where the phyllosilicates were occurring. The rocks appeared a little different from the broken and distorted rocks making up the top of the crater rim, but nothing that shouted, "Here lies clay." So Opportunity did a roundabout walk of the area as a survey and returned to the starting point. This is a strategy often used in field geology where a bit more information is needed to sort the signal from the noise of the rocks. In customary fashion for Coyote Mars, of course, the first leg of the walkabout had encountered some odd outcrops that later proved to be what we were looking for, and it was to these that we returned for a closer look.

The outcrop in question was odd because it consisted of a grid-like pattern of lighter colored rock enclosing darker rock forming a sort of boxwork. On Earth this type of pattern frequently occurs when water circulates along rectilinear fractures common in dense rock, and the margins of the cracks get altered if the water is present for sufficiently long enough, or if it is particularly corrosive. When we analyzed the chemical composition of the rock next to the fractures, bingo, all the elements typical of phyllosilicates were present and accounted for. And further examination in the vicinity showed us that this rock mass was likely something that lay below the rim ejecta and therefore was probably part of the original country rock before the formation of Endeavour Crater. Because that event was long ago during the time of potentially wet Mars it appeared likely that the rocks we were seeing were witnesses to that ancient wet Mars. They had been nicely preserved under the ejecta until erosion finally brought them to light. We were finding evidence of the former presence of water, and lots of it, everywhere.

Later orbital remote sensing told us that phyllosilicates occurred in Marathon Valley farther south along the rim. When we visited that valley, the evidence was less conclusive, but we would find somewhat similar altered older rocks making up the valley floor, where it was again exposing the pre-impact substrate.

GYPSUM VEINS

IT WAS STARTING TO FEEL AS though you could not drive across a terrain dating from the ancient early history of Mars without tripping over some rock record of past water, and lots of it. When we first arrived at Endeavour Crater, we noticed peculiar white lines or veins running across the ground. When we finally came across one that was sufficiently big enough, a couple of centimeters wide and at least a meter long, we analyzed it. The results were a bit shocking. As closely as we could tell from the composition the white veins were gypsum, a type of hydrated calcium sulfate, which commonly forms veins of similar dimension and appearance on Earth.

RIGHT: A vein of gypsum about two centimeters wide and forty-five centimeters long discovered by Opportunity on the rim of Endeavour Crater on sol 2,769.

The thing about gypsum veins is that they form when a lot of mineral-rich water flows through fractures and deposits gypsum along the walls of a fracture through which the water is moving. Over time the gypsum effectively grows outward as it deposits along the walls and the vein widens. The result is a nice thick mass of a mineral that really only forms in that association with water flowing through fractures in rocks. In this case the veins occurred in rock that was just below the contact with the sulfate sandstones of Meridiani Planum, where they had previously overlain the ancient eroded slopes of Endeavour Crater. So one hypothesis that made sense was that because the sulfate-rich Meridiani sandstones had been shown to be soaked in groundwater for long periods, it was possible that that groundwater dissolved, carried some of the sulfates in solution, and passed through surrounding rocks, thus leaving behind gypsum. This was essentially a freebie from Coyote Mars, although the chuckle was the eighty sols that elapsed after we first saw the veins on sol 2,681 before we finally took the bait and got around to checking out one of the veins on sol 2,760. The evidence for former water-soaked ground was racking up on the scoreboard.

HOT SPRING DEPOSITS?

ON SOL 776 SPIRIT WAS PASSING down a small valley on the east side of Home Plate. While we were eager to spend some time looking at the odd Home Plate feature, we had to do a side trip first to find a place for winter, after which we would return and commence its exploration. On its way down the valley it crossed over a small ridge in the center of the valley floor. And after the winter was over, Spirit backtracked along the same path. On sol 1,116, just after crossing the same ridge, the panorama looking back along the path revealed that the small ridge was actually an isolated layer of rather crusty or nodular material. Where the wheels had crushed the material, it appeared somewhat light toned. But we had bigger plans on Home Plate and continued back up the valley.

Now it was just prior to that previous winter that Spirit finally lost the use of the front-right-wheel motor that had been plagued with high motor currents for a long time. As a result, when driving back up the valley Spirit had to drive "backward," dragging the dead wheel like a lame foot. This dragged wheel proved to be an excellent trenching tool because a bit farther up the valley, another look behind us revealed something unexpected. The wheel had inadvertently trenched through some extremely bright, almost white soils. This was a bit too much to ignore, so a short investigation was done on these bright soils, and they were discovered to consist of almost pure silica. Using the spectral tools of the mini-TES, we discovered the best match in terms of mineral type was a kind of opaline silica, not gem-type opal, but a fine microcrystalline type. In fact, the spectra matched that of siliceous sinter, that is associated with hot spring deposits on Earth. Two and two were put together: Home Plate had originally appeared to be some kind of volcanic feature, and hot springs with siliceous sinter are associated with volcanic areas on Earth, so might it be possible that this white material was the remains of a hot spring deposit? The implications were rather stunning. Hot springs are places where there is lots of energy and nutrients for happy little microbes. And if there were hot springs at some time in the past, then if there was a place where ancient Mars microbes would have been happy, that would be a good place.

Steve Ruff was the principal driver of investigations into these deposits and went on to do a number of studies of the occurrence of similar deposits in hot spring areas of Earth. His main point was the fact that there appeared to be a microbial influence on generating deposits of this morphology on Earth and that the similarity is great enough to warrant consideration for future missions. In fact, he led a team of people proposing the Home Plate area as a possible site for the Mars 2020 mission (Perseverance rover), a site that ultimately was not selected.

Ray Arvidson, MER deputy principal scientist, was more cautious and felt that another way to generate high silica deposits was leaching. Given all the acidic water that was apparently prevalent on early Mars, he felt that acidic water percolating down through the surface over the years could leach silica from overlying materials and deposit the silica lower down in the loose soils below the surface. Some experiments looking in detail at the chemical composition in a small cross section of deep soils near Spirit's final site showed that this was possible. The process certainly makes sense. In fact it is a similar process that creates true gem-quality opal on Earth. The conditions under which it forms is certainly Mars-like in many respects: dry environments with water percolating through rocks in a quiet setting and building up silica deposits at molecular levels of accumulation over time. By chance, we had an exhibit about Australian opal at my museum during this period. Jayne was the staff curator, and we both learned a great deal about gem-quality opal. In the end I was coming to think that Mars could be the future site of some spectacular gem-quality opals. We will just need to wait and see. But don't forget—you heard the prediction here first!

The efforts to understand high-silica deposits resulted in a series of outcrop examinations on the side of the valley where similar materials with high silica were present in the form of those strange knobby outcrops like those seen on that ridge a bit farther back down the valley. Spirit continued to churn up more examples of the bright soil as it made its way around the north end of Home Plate on its journey of destiny with a deep dust pit not too many sols later. It was during the investigation of one of those outcrops that the mission experienced the first major global dust storm. But that is another story. Without a doubt Coyote Mars wanted us to work a little harder to get the results on this important material.

FIRST FIELD GEOLOGIC MAPS ON ANOTHER PLANET

THROUGHOUT THE MISSION ONE OF MY pet projects was to keep track of all the different rock types and outcrops in a geologic map made directly from rover observations along the path of rover traverses out to the limits of visibility in the panoramas. The process seemed natural to me because geologic mapping in the field here on Earth proceeds by doing pretty much what we were doing on Mars. You walk along visiting outcrops, inspect rocks and identify different types, and map the contact between the types on a base map. It is all about actually touching, examining, and identifying rock types in "hand specimen" on the ground. You end up with a map that shows the distribution and age relations between rock units along the traverse.

> "The most promising words ever written on the maps of human knowledge are terra incognita–unknown territory."
>
> —DANIEL J. BOORSTIN, *The Discoverers: A History of Man's Search to Know His World and Himself*

The big difference is that on Earth we can cover lots of ground on foot very quickly, can crisscross back and forth across terrain, and fill in this information over a broad area, usually a rectangular area or quadrangle. With a rover the traverses are a bit more limited and the mapping could only take place in a narrow strip to either side of the traverse path. The final maps were long noodles or geologic strip maps. Ultimately the geology along each rover's path was mapped in strips that followed their roundabout paths for many kilometers but were only forty meters wide.

Of course we had been making "geologic" maps of Mars for many years, but those maps were what are called photogeologic maps in which areas with similar morphology as seen in orbital images of the surface were identified, and the different morphologic units were then interpreted to be this or that type of rock. The interpretation was often based on the apparent emplacement process, such as lava flows or layered sediments. So it was all about making educated guesses from the appearance or morphology of the terrain using the big-picture view of the surface from high above. Mapping on the ground, however, takes this to the next level in which you can actually determine from inspection of the actual outcrops what the rocks are. That is the way we do it in the field on Earth. It turned out that these maps I had been assembling from ground observations with the rovers were something new to planetary geologic mapping.

For the first half of Spirit's and Opportunity's missions it so happened that the terrains were pretty much made up of one type of rock. On Spirit it was just basalt of the Gusev Crater floor where it landed. And for Opportunity it was that vast parking lot of monotonic sulfate sands where it spent much of its early mission. So during the first segments of their traverses neither rover was driving in geologic terrains varied enough for any geologic mapping that really revealed anything that was not obvious. But when Spirit reached the varied geology of the Columbia Hills and Opportunity arrived at the complex terrain of Endeavour Crater, the sheer geologic diversity made true geologic field mapping possible. The scale of the units that could be mapped was also much smaller than that in previous photogeologic mapping. For years we had been accustomed to mapping big geologic units, hundreds of kilometers across on the Moon and Mars using orbital images. The images were so big in scope that things like individual outcrops were largely unresolved. But we could see the big characteristics. We had to some degree come to expect that things were so covered with debris and dust that mapping at the scale of outcrops a few meters across, that is, things that occurred at human scales, would not be doable. Here we were finally looking at things on the ground, and the geology was every bit as complex as it was on Earth.

There was yet another thing that made the new field geologic mapping on Mars different. Our previous experience with places like the Moon told us that without erosion similar to the erosion we have on Earth we were likely only to see the messy surface expression of things on other planets covered with debris. Erosion on Earth can plane off kilometers of rock and expose the complex underpinnings of geologic rock masses. But Mars has an atmosphere, and that atmosphere was by all our reckoning considerably capable of erosion in the past. And erosion had obviously occurred, exposing the skeleton of rock structure just like on Earth. We could see outcrops in geologic sections, including the contacts between different rock types laid down at different times. We could do field geology on the red planet at the same human scale of observation that many of us have done on Earth too.

BELOW: This photo shows part of one of the first field geologic maps made from Opportunity observations. The photo was taken along its traverse out to a distance of twenty meters on both sides of the rover as it drove down Marathon Valley toward the floor of Endeavour Crater and then back out to the crater rim. Different colors represent different rock types identified from the rover along the rim of Endeavour Crater. Thin lines over the MRO/HiRISE base image are one-meter elevations.

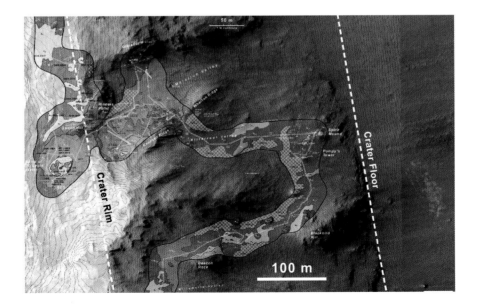

Coming from a field geology background, it seemed to me that Earth-style field mapping was the natural thing to do with our direct examination of rocks and outcrops, and so that is what I did. This type of mapping achieves an unparalleled technical advantage and context information in unraveling Martian geologic stories.

Fellow science team member and former Mars Pathfinder principal scientist Matt Golombek appeared to be one of the few people to understand the strength of this new technique to planetary geology, even commenting once that the rest of the science team probably thought that I was "doing some sort of voodoo" to come up with these maps. There are perhaps many reasons for this. One reason may be that planetary geologists have been used to seeing geologic maps of planets made through photogeologic means, and therefore any map of another planet must surely be some kind of photogeologic map, particularly when the map is presented on a base map similar to the orbital images that have always been the canvas for photogeologic mapping.

Another reason may be that field geologists are few and far between in planetary science. It takes a lifetime to gain experience in planetary science, and that leaves very little professional time to do the time-consuming process of actual field geologic projects. And most of the science team was directly involved with one of the instruments and the process of getting numbers from observations made by that instrument. Geologic mapping was not something that you could measure with an instrument and plot. This meant that the process of how you "do field geologic mapping" was not widely understood in direct terms.

It is not that measuring things with instruments to map out rock types has not been attempted. There have been efforts to use the spectral characteristics as seen from orbit and the chemical characteristics of the landscape as measured in visited outcrops to map out similar and different rock units. But there are challenges to making sense of the results with instrument measurements of that sort. The difficulty with that approach is that the surface of Mars in older terrains is incredibly altered chemically and the alteration manifests itself randomly. One outcrop can be chemically altered by some event in the deep past, and the next of the same original rock type is entirely different. Sometimes the chemical data appeared unrelated from site to site. So frequently what I mapped on the basis of the usual field geologic hand specimen characteristics as the same unit would often appear somewhat distinct in the chemical analyses.

I labored on throughout the mission, doing my voodoo field geologic mapping on another planet. But the maps resulted in some interesting insights. One advantage of keeping track of different rock units from site to site is the ability to correlate to outcrops as being essentially the same geologic unit laid down at the same time. In the end, an extensive stratigraphic column could be made showing how all the units were related in a time series of stacking and how all the analyses fit into the geologic history recorded by this stratigraphic column.

The final map along the entire traverse serves as a type of record of all the distinct rock units visited and can be useful when later attempting to correlate the results from site to site. The results were another mission first, the first field geologic maps done on the surface of another planet. In the future, the significance may filter down into the collective conscience of the review of Mars exploration history. Anyway, that is my interpretation and I am sticking to it.

THE NEAR-DEATH EXPERIENCES

AFTER MANY YEARS OPERATING ON MARS, both rovers were showing signs of age. Every year at the science team meeting, John Callas gave us a summary of the current status of both rovers as time passed. John was the project manager working for the Jet Propulsion Laboratory in the capacity of administering both the science and engineering team involvement in the mission. John was an organized and even-handed fellow with a radio-announcer voice who had the task of herding all us scientists through the yearly cycles of budgets while working with the engineering team to make sure they had what they needed to keep the rovers operating successfully and fulfilling mission goals.

During these yearly presentations, John would usually start with a review of what part of each spacecraft was showing wear and what the engineering work-around had proven effective. Because both rovers were operating far beyond their designed ninety-sol lifetimes, things were bound to get worn out. On a side note, early in the mission, toward the end of the ninety-sol nominal mission, we were complaining that the office furniture in the science team area of Building 264 was wearing out and breaking while the rovers were remaining in perfect health.

The standard sort of statements that arise in this type of situation began going about the team, such as: "If we can put a rover on Mars, we should be able to design a chair that doesn't break after a few months," or "I can't even drive my car for three months without some sort of maintenance, but the rovers are on a freezing, dusty planet, and they are fine."

The list of ailments on the rovers grew with time. For example, one of the instruments, the Mössbauer spectrometer that measured the complex chemistry of iron, was powered by a small radioactive source, cobalt-57, that has a half-life of 271 days. After one thousand days, not much of the source material was left and so that instrument on both rovers could not operate effectively. That was just a consequence of living too long since the original plan was a mission that was expected to be much shorter than that.

Similarly, the dust collected on the optics of the mini-TES thermal-emission spectrometer early too, and that instrument was lost to us. All we had left was a chemical element analyzer, a microscopic camera, and a RAT brush. The RAT grinder on Spirit had worn out just after sol 419. Concerns for wearing out the RAT grinder on Opportunity were such that we avoided using it in order to have it available for some particular important find. The cameras fortunately continued operating.

> "Death when unmasked shows us a friendly face and is a terror only at a distance."
>
> —OLIVER GOLDSMITH,
> *Threnoda Agustalis: Sacred to the Memory of Her Late Royal Highness the Princess Dowager of Wales*

OPPORTUNITY GETS ARTHRITIS AND HAS MEMORY LOSS

EARLY IN OPPORTUNITY'S MISSION, IN AN uncharacteristic bit of bad luck for that rover, it was observed that the batteries were quickly discharging after the rover went to sleep overnight. It turns out that a small heater used to warm the arm joint motors was stuck in the "on" position. The only way to turn it off was to completely shut down the rover at night. Worse still, as Opportunity was doing its death march across the plains heading south to Victoria Crater and beyond to Endeavour Crater, the "shoulder" joint on the arm stalled.

After a lengthy investigation it was decided that perhaps the thermal cycles from the constantly "on" heater followed by the intense cold of night when the heater was turned off could have led to a failure of that motor. To avoid having the arm stuck in some inconvenient position that would interfere with using the arm at all it was decided to avoid stowing the arm and leave the arm hanging out in what became known as the fisher-stow position. Thus, Opportunity drove around during its remaining years with the arm dangling out like it was on a fishing trip.

Then there was the problem with rover memory. Part of the memory was similar to the flash memory of a typical thumb drive that many of us use for transferring data from one computer to another. This was the memory that stored all the science and other data collected on a given sol. Like any flash drive the data stayed on it even when power was turned off. So daily observations that could not be downlinked were generally stored in the flash memory. But as Opportunity was working its way along the rim of Endeavour Crater, there began a series of anomalies in which Opportunity would wake up some mornings with a kind of amnesia. Spirit had begun to experience somewhat similar symptoms late in its mission.

Ultimately the problem on Opportunity was tracked to a sketchy flash drive memory. Why the problem arose was never really resolved. But the only solution was to avoid using flash memory. But the problem with that was that the other type of memory that we could use instead was the kind that erases when you turn off the power, such as when Opportunity shut down at night because of the heater motor that would otherwise drain the battery. This meant that the final year or two of its mission required that we downlink anything we wanted, such as the entire sol's activity results, before each night. And because the amount of data that could be downlinked was variable and generally smaller than what we could collect in a day, the plans going forth required that we only collect the data that we could get from the predicted downlink on that sol. But it worked, and I have to say the work-around streamlined the wait for data. There were some observations earlier in the mission that would sit on the rover for days because they were not high priority and only came down to Earth after considerable time.

There were other failures on Opportunity, such as a steering motor on the front wheels that left it steering sideways for a while until it suddenly popped back after an attempt to straighten it one day. After that we avoided using that steering motor and just did tank-like turns when necessary. The next time it could get stuck permanently. We were beginning to expect that we would wake up one morning and find Opportunity using a walker.

SPIRIT GETS A "FLAT TIRE"

EARLY IN SPIRIT'S MISSION THERE WERE signs of excessive current on the motor driving the front-right wheel. Each wheel had a separate motor, much like any modern electric vehicle, and the front-right wheel currents were alarming. High current in a motor is usually an indication of impending failure. Given the terrain of Mars, we needed all wheels operating. As a result the engineering team did many types of mitigation measures, but the most visible activity was driving "backward," although technically neither direction was a forward on the rovers, in order to take the stress off the front wheels. Then there was driving backward with the front-right wheel operating only 10 percent of the time to reduce the number of cycles.

Ultimately the front-right wheel quit working. But, of course, Coyote Mars thought it would be particularly funny if it happened when having all the wheels functioning was particularly critical. And so it was. Spirit was driving across a valley near Home Plate on its way to a hillside where it could operate tilted toward the low winter Sun as was our custom. Time was running out, as fall was coming up quickly, when we really needed to get to those slopes. But just as it was crossing the valley to that hillside it became mired in the soft sand and fluffy dust lying in the valley bottom. On the next sol an attempt to back out was unceremoniously thwarted when the front-right-wheel motor stopped working. Now we were in for it. Driving with five wheels was bad enough, but while stuck in a sand drift? Come on! After several sols we managed to wiggle Spirit free and hightailed it to the closest place where we could tilt the rover for winter driving backward and dragging the front-right wheel like a zombie with a dead foot. And this was none too soon. Another sol or two and there would have been insufficient power to keep the rover alive. Spirit was ever the drama queen that way.

Coyote Mars was having a particularly good time watching us bite our nails on that one. But like all his tricks, what looked bad turned out to be just a way to make us dig a little deeper, so to speak, for an answer and a dramatic result. On leaving the winter site the next spring, dragging the front-right wheel was like a continuous trenching tool. And it was in this trench that we churned up one of the more fascinating discoveries of the mission, the strange bright silica-rich soils that I mentioned earlier, that were either hot spring deposits or some spectacular rock leaching associated with the local volcanism. Without the bum wheel we may never have seen those deposits.

DUST DEVILS

YOU NEVER KNEW WHEN IT WOULD happen, but you could count on that gust of solar-panel-cleaning wind happening sometime during the windy season. It was just that with wind also came more dust if the winds started getting out of hand. A curious visible evidence of the movement of dust was the occurrence of dust devils, as they are called in the southwest United States, or dusty whirlwinds.

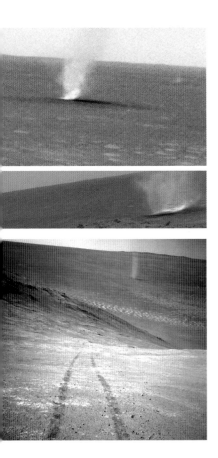

Images of dust devils crossing the Gusev Crater floor plains as seen by Spirit from its perch on the slopes of the Columbia Hills on sol 486 (**TOP;** colorized) and sol 456 (**MIDDLE**).

BOTTOM: View of a dust devil moving across the floor of Endeavour Crater captured while taking an end-of-drive Navcam panorama. This was just after a drive up Marathon Valley by Opportunity on sol 4,332.

The dust devils of Mars were first observed by Spirit, and they were ghostly things. Spirit had landed in a particularly dusty site, and there was evidence for dust devils everywhere. An interesting effect of dust devils that had been observed from orbit for some time were curious dark curved and straight lines that etched the surface of Mars.

The lines were originally interpreted as dust devil trails by John Grant, when he worked at Brown University. Because dust covers everything on Mars, when a dust devil passes over an area it effectively vacuums up the local dust, revealing the true darker surface beneath.

During the summer months dust devil activity picks up as the Sun heats the ground and local convection starts lifting the thin air and swirling. We knew that dust devils were prevalent in the area where Spirit landed based on all the dark streaks seen in orbital images, so to see if we could capture one of these dust devils in action Ron Greeley, the Martian windblown sand expert, and his students set up a series of observations to capture one in the act. Spirit had a perfect observation platform for doing so as it began its ascent of the Columbia Hills. Every afternoon, over a period of time, Spirit was commanded to take a series of Navcam images looking west out over the Gusev Crater floor around 2:00 p.m. local time, the time when the surface heating is at a maximum. Sure enough, in a series of spectacular images the ghostly shapes were caught dancing across the surface. We were even able to take a series of "movies" by taking a series of Navcam images every few seconds during the magic-afternoon time period when they were most active. The

movies show ephemeral columns of swirling dust moving rapidly across the landscape, and even leaving dark streaks in their tracks. There was something eerie about seeing something move on Mars, like some strange life-form that was born out of thin air like some ghostly apparition, and then just as quickly disappearing.

Dust devils were less common at Opportunity's site. But perhaps the most spectacular example for Opportunity occurred one afternoon as it was climbing up a slope in Marathon Valley on the rim of the twenty-two-kilometer-diameter Endeavour Crater. At the end of the drive, as the standard Navcam panorama was being acquired for use in planning the next sol's activities, a towering dust devil was caught dancing across the crater floor in the distance. As chance would have it, the dust devil was beautifully framed along with Opportunity's tracks on the slope of the crater rim. There is something about this image that records the alien experience of working on the red planet.

DUST STORMS AND THE WINDS OF MARS

DUST STORMS ARE ONE OF THE big hazards of working on Mars. It is not that the winds are a hazard, as some fiction stories would have you believe. The Martian air is almost one-hundredth as dense as Earth's air. You can calculate the force of a wind gust by taking into account its density and speed, and the results show that a typical wind on Earth that would be at near damaging speeds, say one-hundred kilometers per hour, would feel like a gentle breeze on Mars. The thing that makes a Martian dust storm a thing to be dreaded is the fact that when they occur the atmosphere can be nearly opaque with dust. If you are a solar-powered rover, the drop in solar energy when a dust storm is in full swing can be deadly.

Fortunately there are regional dust storms only in summer months, and only every few years does a dust storm spawned in some region suddenly blossom into a global event. A global dust storm can quickly turn the entire planet into an opaque dusty globe as seen even from Earth. In fact the current Martian calendar, which we shall explore shortly, begins with the first observations of a global dust storm event in 1956. This was Martian year one and in 2021 we are in Mars year thirty-six. Another infamous global dust storm of course occurred as Mariner 9 went into orbit in 1971.

We as yet do not know exactly why some regional storms die out shortly after they develop while others become global. We are learning with each new experience, but Coyote Mars wants us to do a little more research before revealing any of the secrets.

Unfortunately for Spirit and Opportunity, a global dust storm occurred just a few years into their missions in 2007 and the lives of both rovers were correspondingly imperiled at the time. Spirit had just discovered the remarkable white soils near Home Plate and around sol 1,261 was attending to a rock named Innocent Bystander because it was accidentally crushed, exposing a bright interior; while working on an outcrop that would provide some further information on that remarkable discovery, the first global storm hit. I was long-term planning lead on those sols and the low power activities came to a screeching halt.

To understand how intense this dust storm event was we can look at the power levels and atmospheric dust plots for Spirit over the course of its mission. The power levels and tau (a measure of how dusty the air is) are like roller coasters. The power goes up to a peak in midsummer and is followed by a long slide to a trough-like low in winter. The tau, or atmospheric dust, goes down in winter to a value of 0.5 and periodically peaks when there is a local dust storm, and it peaked at its highest extent during the great global dust storm of 2007 when tau reached an outrageous value of 4. Not long after the peak, the storm lessened and operations had resumed, though all that dust in the atmosphere began to rain down, and the dust factor (a measure of solar panel power output) that dropped with the increased dustiness of the solar panels began to slowly but surely slide down. Fortunately Spirit had one of those solar-panel-cleaning events a few sols before the global dust storm began. But as always, what the Martian winds giveth, they may also taketh away.

Meanwhile, during the same global dust storm, Opportunity was sitting on the rim of Victoria Crater around its sol 1,230. Recall from the discussion above, we measured atmospheric dust and opacity with a value called tau, and during the storm tau was a value of 5 where Opportunity was; in normal Martian skies during the clear winter months, tau is 0.5 and only gets to 1.0 during spring winds. During a global storm the drop in sunlight turns the sky dark and the landscape takes on a late twilight appearance. Opportunity actually recorded this ever-decreasing lighting during a remarkable series of images taken during the dust storm between sols 1,205 and 1,235.

The only way both rovers could survive was to reduce activity to a bare minimum, using what little light there was to recharge the batteries as best they could and hope for an early end to the global storm before power dropped below survivable levels. Had the storm lasted longer or become more intense, then the missions may have met an earlier end. As it was, the skies cleared after a few weeks and operations resumed, and both rovers went on to explore for several more years. But the storm simply reminded us that you could never assume that tomorrow would be as good as today. Coyote Mars was ever on the lookout for an opportunity to do mischief.

AND DEATH EXPERIENCES

HEARTBREAK IN WESTERN VALLEY

AROUND SOL 1,371 SPIRIT WAS MAKING its way north along the west edge of Home Plate on a roundabout exploration when it tried to cross a little depression, informally named Tantalus Crater, roughly nine meters across, that was probably an old and very degraded small-impact crater. Coyote Mars thought that the name was too good an opportunity to pass up for a little irony. As Spirit attempted to climb the other side it began experiencing considerable wheel slippage. Apparently, the interior was coated with some particularly deep loose soils and traction was just not available to climb the other side. As usual we were on our way north to find a winter perch facing the Sun on the north edge of Home Plate and did not have a lot of time to do so before the onset of winter and the corresponding drop in solar power.

After several sols of slipping and sliding, the rover engineers decided the best bet would be to turn around and follow a narrow strip of more solid ground on the western edge of the depression. Ultimately Spirit was able to make it back to solid ground, but in the process came very close to the edge of the escarpment on the west side of Home Plate, which of course was cause for a little tension. At that point it was overlooking

what we had dubbed the Western Valley, a long north-south trough that separated Home Plate from an adjacent rocky ridge on the west side of the valley.

The science team had wanted to eventually work our way around the north side of Home Plate and enter that valley as a way of continuing south to other features of interest while at the same time perhaps getting a good look at the outcrops that would likely be exposed on the western side of Home Plate. And the perch there on the edge at Tantalus Crater gave us a good overview of the valley. But the engineers were against that plan because the valley looked a little rocky and treacherous with more loose soils. So the strategic plan was to go south by way of the top of Home Plate after the winter, over on the north edge. Little did we know at the time that we would

View south from Spirit's final location on sol 2,114 of the enticing landscape that was to be the next major stop in its explorations.

end up driving through that valley after all. The north edge of Home Plate proved too slippery to ascend again with just five active wheels. Spirit gave up the attempt and eventually headed south down that valley as the only option despite many attempts by the engineers to find an alternative. It was shortly after we made the decision to go south through the valley, as we were working our way around the north side of Home Plate and driving west to the head of the valley, that Spirit again started to have flash memory problems in what was best described as "amnesia," a problem similar to that experienced by Opportunity in its final years. Spirit was getting old.

But the engineering team was right. As we started to move down the Western Valley and were about halfway through it, on sol 1,892, May 1, 2009, Spirit was driving backward, as was the custom, dragging the dead front-right wheel. During this drive it attempted to cross a patch of soil with just a few rocks. Now, normally, small rocks on the surface indicate that solid ground was probably present and very little soil. But it was a trap. The rocks were apparently sitting on top of a crust that was hiding a very deep and fine dusty hollow. Spirit broke through the crust and buried the full operational driving wheels on the west side up to the hubs, stalled out, and awaited further commands.

Examination with the cameras showed that the soils were a particularly powdery fluff. Only the east side had any hope for traction, but that was the side with one dead wheel. There were only two wheels with limited traction, and that proved insufficient to extract the rover. In its temporarily stationary position, it was hoped, Spirit had a "tantalizing" view of the scenery to the south where a large butte beckoned of even more surprises, if only Spirit could get there.

After a lengthy period in which many simulations were done using the test rover at JPL to determine the best plan for extraction, in November 2009 many attempts were commenced to move. But with only two wheels with minimal traction, nothing was working. During this time, the engineering team also suspected that a large rock that Spirit had driven over was possibly high centering on the belly of the rover, and that driving might be further hampered by this problem as well. And there was also the fear that the rather weak belly pan could be ripped by the rock if we got too vigorous.

In order to see if the rock was there and if Spirit was high-centered, we decided to do something we had not done before. Since we could not see underneath us, we realized that we had a sort of "inspection camera" on board in the form of the Microscopic Imager on the end of the arm. Maybe the Microscopic Imager could be lowered down, looking back underneath the rover, and we could then determine how close the rock was to touching the rover underside. The problem was that the Microscopic

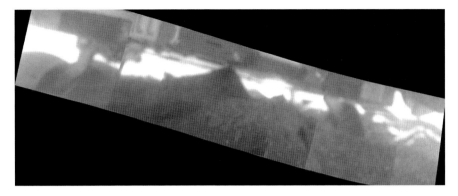

ABOVE: One of the first selfies made by Spirit on sol 1,925 using the Microscopic Imager on the end of the rover arm to investigate the proximity of a large rock threatening the underside while Spirit was trapped in deep sand at its last location.

Imager was not designed to focus on anything beyond a few millimeters at most and any images would be severely out of focus. Nonetheless it was felt that the view, however out of focus, would be better than no information at all.

Thus, Spirit took its first kind of selfie with a series of Microscopic Imager frames stitched together. There was indeed a rock there, and it was a pointy rock aimed right at the belly. But we decided to continue with attempts to move, and with luck maybe progress could be made that would elevate the rover away from the rock.

By this time winter was coming up fast and, with it, reduced power. Eventually there would be insufficient power to make continued efforts to drive out of this mess. We effectively put the pedal to the metal, and slowly but surely over a couple of drives, the rover was starting to scramble out and skittered a bit to the left; the rock was emerging from underneath and hope was high that we would get unstuck.

Then tragedy struck. On sol 2,097, November 28, 2009, the rear-right-wheel motor died. The one wheel with any hope of traction, the one on the upslope side, was out of action. Spirit was effectively a four-wheel drive, and two dead wheels were on the same side. Spirit was not giving up without a fight, and to show it, the long-dead front-right wheel began working at about quarter capacity. But it was futile. Any future mobility at that point was going to be severely restricted. But wait, there's more. The last drive was essentially the limit of available power and there was no more energy at that advanced stage of the season for driving. And with only four wheels that was unlikely to be happening in the current situation.

The bells were tolling. Spirit was not in a tilt position that we needed for winter power, and there was no way to drive into a tilted position. The next several months

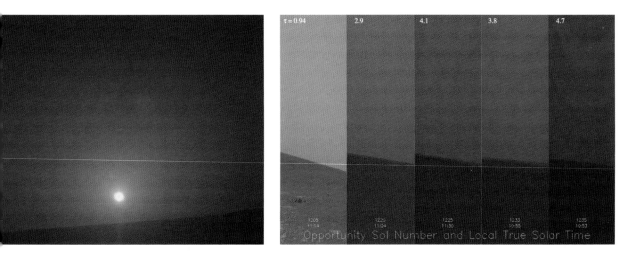

τ = 0.94   2.9   4.1   3.8   4.7

LEFT: Image of the sunrise over the eastern rim of Endeavour Crater on Opportunity's sol 4,999. At this, its last location, it had traveled 45.1 kilometers over the surface of Mars and was about to celebrate its five thousandth sol on Mars. **RIGHT:** Composite of images taken by Opportunity showing the decrease in sunlight over 30 sols during the previous great dust storm of 2007. The numbers across the top indicate the increasing dust in the atmosphere.

then were little more than a deathwatch. The last communication with Spirit with the onset of winter was on sol 2,188, March 22, 2010. Presumably power had dropped to such an extent that the internal clock had quit. It was hoped that perhaps with the onset of the summer in March 2011 that Spirit might again have sufficient power to reboot and begin communications. Efforts to receive a response were made throughout the Martian spring and early summer. But no response was forthcoming. Spirit's heart had finally given out. Coyote Mars had conspired with that one additional dead wheel, perhaps somewhat mercifully, to give Spirit a certain but heroic end.

DEATH COMES FOR THE OPPORTUNITY ROVER

ON SOL 4,999, OPPORTUNITY WOKE EARLY in its site within Perseverance Valley on the inner slopes of Endeavour Crater and took a beautiful image of the sunrise in honor of its coming five thousandth sol on Mars. Times were good. Opportunity even celebrated its five thousandth sol on Mars with its first selfie (actually acquired over

the period between sol 4,998 and 5,004), taken by holding the Microscopic Imager out in front, pointing it at itself, and moving it through a sequence of positions that ultimately yielded the first actual look at the rover since it landed. Someone from the Pancam team proclaimed, "Why, yes, the Pancam mast is actually still there!"

But the celebratory sunrise image and selfie were in a way perhaps Opportunity's poignant way of saying a final goodbye. A few sols later, as it continued to work on an outcrop of unusual basaltic rocks in the valley floor, the alarm went out during one of our SOWG uplink meetings that a significant dust storm was brewing to the northwest of the rover site. The intensity and rapid growth suggested that this might be another "big one" similar to the global dust storm that both rovers had weathered in 2007.

We had grown accustomed to the periodic call of "Wolf!" every summer as the atmosphere got dusty and regional dust storms blossomed here and there on the planet. But this one proved to be bigger. It was one of the largest global Martian dust storms since the beginning of space exploration. And it developed into a full-blown event right over the rover.

The decreasing light and disappearing Sun were simulated in a series of images that highlighted how dire things had gotten up to the point on sol 5,111, when we last communicated with Opportunity. In that last downlink from Opportunity, on sol 5,111, June 10, 2018, nearly fourteen and a half years after landing in Meridiani Planum in January 2004, Opportunity posted a low energy of 22 watt-hours and a tau of 10.8, the highest tau we had ever measured. While it was expected that the storm would abate in a few weeks and the returning Sun might allow Opportunity to gain sufficient power to start communicating, we never heard from Opportunity again, despite attempts to contact it over the next eight months.

I felt that the final command event was a little morbid and way too sad, and I could not bring myself to attend it.

We all gathered at JPL on February 13, 2019, for the formal announcement of mission end by the NASA administrator and associate administrator and a few reminiscences. On the evening of the final attempt to contact Opportunity everyone was going to gather in the control room at JPL to watch the final command. Of course it was a typical dramatic funeral with a cold rain driving us into the auditorium instead of the planned outdoor activities on the JPL mall. There were many sad faces. But we all agreed that we had had a hell of a ride. We had a final science team meeting at Caltech in Pasadena in June 2019, during which we delivered some final reports and we saw each other for the last time.

ANCIENT, WET MARS, I PRESUME?

WE WENT TO MARS LOOKING FOR evidence of its early climate using tools similar to those a geologist uses to look at rocks in the field to determine how the rocks formed and then interpret the geologic history of rocks. Did we "find water" on Mars? Yes, we did. Of course we did. But by the time we had found good evidence for water in the past as recorded right in the rock outcrops in front of both rovers, we had found water or evidence for lots of past water many times and in many forms from orbit and now from the ground. And like Stanley looking for a Livingstone, while the journey was a different one, the Mars Exploration Rovers might easily have said, "Ancient wet Mars, I thank God I have been permitted to see you!" like Stanley's second—but not as often quoted as his first—words to Livingstone after years of searching exploration. By the time we found evidence for early wet Mars, we had, like Stanley, already found "rumors" of it many times, uncovered evidence for it repeatedly in our minds. The rovers might easily have said by this point something like, "Ancient, wet Mars, I presume?"

To put it bluntly, it was pretty obvious that water had been there before. When the rocks are so corroded that they have the consistency of an adobe brick at best, then there has probably been a lot of water at some point in the past that enabled that corrosion. When everywhere you look you see minerals that require water to be transported, precipitated, and fixed in the matrix and cracks of every rock, then it does not require elevated astuteness to conclude that there has probably been lots of water.

> "Ah well, that is a missionary's life; to plant where another shall reap."
>
> —WILLA CATHER, *Death Comes for the Archbishop*

But casual observations and descriptive scenarios are not how science works. We had to document it with quantifiable results from the chemistry and geology of the rocks in front of us. That is where all the roving and measuring and looking at rocks came in. That is the part of science that is not often seen in the public reports. In science there is a lot of slogging, measuring, and analyzing. And then, if you have all your ducks in a row and you stand straight and enunciate correctly, there may be a murmur of vague approval; it takes a lot of work and evidence to get more than an approval and instead get something closer to a scientific consensus. We did all that and a lot more.

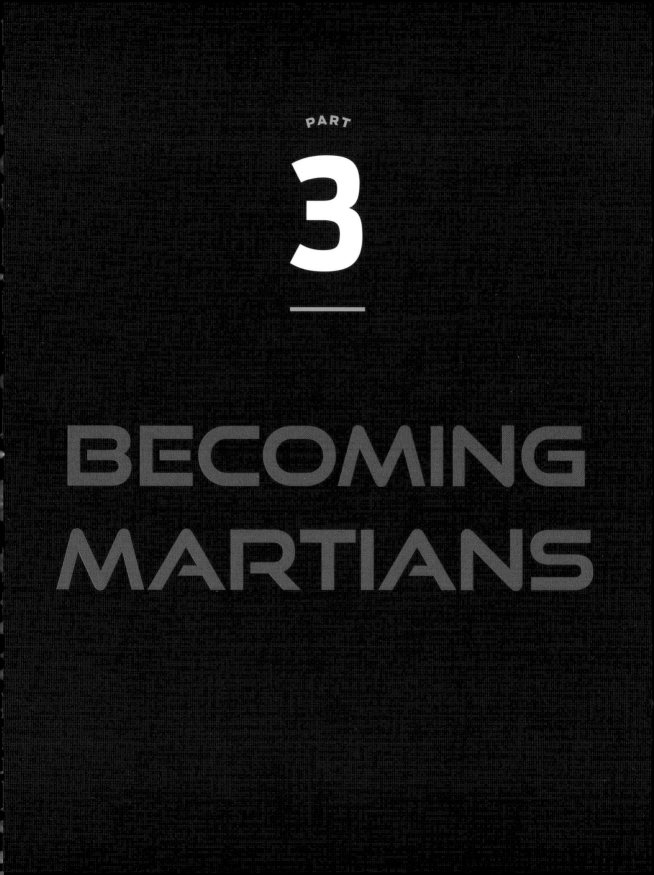

BECOMING MARTIANS

TEN

Experiencing Mars: Living on Another World

"The best and most beautiful things in the world cannot be seen or even touched—they must be felt with the heart."
—ANNE SULLIVAN, PARAPHRASED BY HELEN KELLER, *The Story of My Life*

ONE THING THAT WE LEARNED FROM roving the surface of Mars was the fact that it is a "real place." Before our experiences of several years on the surface of Mars, it was an alien place, seen only from afar. The actual experience of the unexplored alien world was a thing of the imagination, not one of daily life experiences. The Mars we roved was certainly alien, yet we became accustomed to the daily and seasonal rhythm, its starkly pristine landscapes, and the daily lighting. But like any strange new world, there are strange new experiences that can remind you that you are exploring more than just an untouched landscape—there are explorations of new concepts and perceptions as well.

While we had grown accustomed to the stark landscape from our orbital views over the previous decades, to really experience a place you must get down and walk in it, experience daily life on it, feel the passage of each day from sunrise to sunset, look at its sky, and watch the progress of seasons. We had been on the surface before, but with Spirit and Opportunity we had the chance to live on the surface and experience these things in a way never before possible. Mars had become a real place, and an alien place it is.

THE ROBOTIC "WE"

MARS IS CURRENTLY A WORLD OF telepresence. There is a popular meme that says, "Mars is the only planet known to be inhabited solely by robots." Someday, it is likely that there will be people on Mars, but for now we visit Mars through the "eyes" of robots or rovers. This brings up a curious thing that occurred during the missions that is insightful regarding the interaction of humans and machines. For most of the mission we spoke of the rover in a first-person plural context. We used the word "we" and the actual names of the rovers almost interchangeably, as I have tended to do throughout this narrative. It was as though we were on a small bus driving over Mars: "We drove over there. We measured that."

Outside observers picked up on this tendency very early and frequently commented on the team's propensity to speak as though the team itself were on Mars walking around. Individually most of us on the team felt that we *were* on Mars.

Another common theme to this discussion was the statement that "we would go to work on Mars every morning."

There was at least one dissertation done at the time that explored this social phenomena from several angles, including the team dynamics associated with just the process of getting information through telepresence. In a world in which many events, gatherings, and even field trips are taking place on teleconferences, the concept is becoming less alien perhaps. But the real phenomena to me is the fact that once you have immersed yourself in the environment via this type of telepresence the world that you see via remote means becomes very familiar although in a fuzzy way, somewhat like looking at everything through a tunnel. The mind builds up a mental map of your setting, but the eye can only experience a small and very narrow scene.

To add to the vagueness, the environment that you see is sampled very infrequently, when the downlinks occur. So that mental map gets accumulated with time rather than in one look. But the mind already has a map of the setting with big blurry spots and the new images are simply plugged into that. It's somewhat like looking through a pair of glasses with clear lenses briefly, and before you can get a complete survey the glasses are muddied. Then you clean them off and get a continued look, but before you complete the survey of your surroundings the glasses get muddy again. It is a sort of herky-jerky, gradually built-up view of the world. But a view of the world nonetheless develops, and you move through it.

Somewhat related is the tendency to anthropomorphize the rovers. Someone once commented that the shape of Spirit and Opportunity was itself responsible with its human height, the long "neck" of the camera mast, and the two camera eyes widely spaced at the top of the mast inducing a kind of cuteness to the presentation. I had somewhat mixed reactions to another tendency, frequently voiced by the engineering team, to give the rovers a gender. Throughout this narrative I have consistently referred to the rovers individually as "it," but members of the engineering team frequently used "she" when speaking about a particular rover. Presumably this was an extension of the old maritime custom of giving ships female names and referring to them as "she."

But there was also a vast fan base that viewed the rovers as individuals, almost like living beings. All this pales in comparison with the heartfelt outpourings of grief when Opportunity died. There were a few discussions of why there was so much emotional attachment to what in the end was a machine that just broke down. The answer appears to be complex, but it has more to do with the human ability to empathize with just about anything. And the more that thing carries the hopes and aspirations of many people, the more it assumes a "presence," or a type of artificial soul not bestowed on other inanimate devices. The list of essays on this topic is long. Perhaps the whole subject represents the emergence of what is a growing field of sociological debate as artificial intelligence becomes more prevalent in our daily lives.

NEW WORLD, NEW CLOCK

TIME ON MARS IS A NEW frontier beyond the physical kind of an untraveled landscape. On Mars the "landscape" of the day is different. One of the unexpected and previously unexperienced hazards of planetary exploration by remote robotic rovers is the problem with the length of the Mars day, or sol. No group in human history, before this mission, has had to work on clock time that is different from our accustomed Earth twenty-four hours, and there are consequences, yet we explored the alien time of Mars thoroughly with the first long-lived surface missions of Spirit and Opportunity.

It was not widely appreciated before the first rover missions such as Pathfinder and the MER missions, but the time difference between Mars and Earth days and years is just enough to set many things askew in our human world accustomed to a twenty-four-hour day and a 365-day year. The Pathfinder mission was short enough that the team controlling it just lived with the consequences of being on a clock that was out of whack with Earth time. But the Mars Exploration Rovers were expected to last at least three months and something needed to be done to accommodate the time differences. It is a new world of interaction, new terminology, and new timepieces.

> "'Tis with our judgments as our watches—none go just alike, yet each believes his own."
>
> —ALEXANDER POPE,
> *An Essay on Criticism*

What happens when the basis of your clock, the Earth day of twenty-four hours, is no longer applicable to the day on the planet where you happen to be? A day, of course, is one rotation of Earth, or the time that elapses from sunrise to sunrise. But the solar day on Mars, the time between successive sunrises, is thirty-nine minutes and thirty-five seconds longer than an Earth day. And that is where the problem starts. With two planets with different-length days, operation of anything jointly between the two can be likened to a gear clash between two gears with a different number of teeth.

Here is how that "small" difference is a problem. Since the Mars day is nearly forty minutes longer than an Earth day, after about fourteen days on Mars, the clock time on Mars has dropped behind that on Earth by around nine hours, or not quite a half a day. This means that if you get up and go to work on Mars time every day, as we did in the first ninety sols of the mission, then after a little over two weeks your Mars clock is behind the Earth clock by nearly half a day. What that *really* means is that after a little over two weeks you are driving to work in end-of-day rush-hour traffic instead of morning rush-hour traffic.

This can happen easily on Earth if you are working third shift or a nighttime job. But in the case of working on Mars, you are going to work at the beginning of the sol, on Mars, and the clash with the diurnal cycle of sunrise and sunset on Earth can be shocking, disorienting, and frustrating. Let's face it, at the end of a "long day working on Mars," which is already forty minutes longer, you can find yourself wanting a nice dinner, but every couple of weeks the only thing happening in the world physically around you is breakfast! You get up to start work on Mars and are ready for breakfast, but the world is sitting down to dinner. And those are the good times when there is an obvious offset. It keeps changing. In between, there are times when you have no idea what the time of day is on Earth.

No group in human history, before this mission, has had to work on clock time that is different from our accustomed Earth twenty-four hours, and there are consequences, yet we explored the alien time of Mars thoroughly with the first long-lived surface missions of Spirit and Opportunity.

The best solution is to just not interact with Earth at all if you can help it. We tried to do this by blocking all the windows in the building at JPL so that we would not get any clues from outside the windows. To all intents and purposes, however, instead of feeling like we were working in a habitat on Mars, we felt like we were working in the dark of night when there was no outside light. It felt like we were doing a perpetual series of "all-nighters," and our sleepiness and overall tiredness accumulated in the space of ninety sols accordingly. Everybody said we looked wasted at the end of the primary mission. This was despite trying to stay on Mars time.

One of the first things that the planning by the team had to grapple with was the length of the day on Mars. But what about daily measures of time? Is a Martian minute or hour the same as an Earth minute or hour? What about clock time on Mars? If a Mars day is slightly longer by thirty-nine minutes and thirty-five seconds, can a regular Earth clock be made to work for Mars? With some mechanical work you can use a clock designed for Earth time. But first you need to slow down an Earth clock by thirty-nine minutes and thirty-five seconds each twenty-four hours, that is about 1.55 minutes added to each hour, in order to make the twenty-four hours of an Earth clock stretch into the length of a Mars sol. Fortunately that is possible with work by a clock expert. In fact many of us had "Mars watches" that we used during the time we spent working on Mars time, and they were designed to do just that. Back in Albuquerque, Jayne wore her Mars watch so that she knew what time to contact me. The Mars watches represented a flexure point in interplanetary history actually. It was the result of the first time that a group of people needed to live and work on the sidereal time of another planet and therefore needed watches and clocks that operated on the daily cycle of another planet.

ABOVE: One of the watches redesigned to run on Mars time. We wore these during the first ninety sols on Mars to keep track of Mars local time for each rover.

Another consideration is the difference between local solar time (LST) and time zone time. So far, all the successful Mars landers and rovers have just used LST because there are no organized time zones on Mars like those we use on Earth to keep whole regions on the same clock time. But with enough equipment operating within a few degrees of latitude and longitude from each other, having a system of Martian time zones might make sense someday. Otherwise two stations within a few hundred kilometers of each other would constantly need to convert between two local solar times if they would have any hope of keeping to a common schedule like radio check-ins. It gets crazier than that. Say you have a mission clock that is set initially for the time at the site of landing, as, say, a rover that lands at one place. And say that that rover travels a great distance to another longitude, a "great distance" in this case being maybe several kilometers or maybe even tens of kilometers. At the end of the traverse the rover's mission clock that was set for the landing site is now incorrect for the local solar time at the end of that long traverse in longitude. How could you time a precise event like a solar transit of Phobos across the Sun or the precise time of sunrise, or, more important, the precise time of earthrise?

We solved some of these problems when we started operating on our normal Earth schedules. But this also brought into play another set of complications. We needed to communicate with each rover once each day, mainly to see what it had achieved by the end of its day. This allowed us to decide what commands we wanted to send up before the next Martian work day. But if you are on Earth time and the

All of this was academic, of course, until a group of people, us, on the first long-lived Mars rover mission had to live and work on Mars time.

rover is on Mars time, its end of sol happens forty minutes later than each day here on Earth. After a little more than two weeks, the rover end of sol is after bedtime here on Earth. So we might not all see the results of the previous sol of rover activity until we get into the office many hours later. By that time there are only a few, if any, hours before the rover must wake up again and start its new day. Very quickly it becomes extremely difficult to review the results of the rover's previous sol, come to a conclusion, and decide what needs to be done on its next day all before it is too late to send the day's list of activities up to the rover.

The only way around the problem, short of working third shift, was to skip a day of direct planning. But in the primary mission, that is, the first ninety sols of the mission, the rover was viewed as a finite resource that had an expiration date, which we couldn't read, and so the safest assumption was to assume that the expiration date was sooner rather than later. So every day counted. Maybe the expiration was the next day! Or maybe not. But it was better to avoid wasting time and having regrets.

Later it became too expensive to have the team operating on this odd sliding shift, so we switched to an Earth normal schedule and lived with the loss of a Mars work day every few days. But there were ways to make this useful because some activities could be automated during the course of routine activities at a given site. For situations where we needed to fill in a day there were things that we could have the rover do on a second sol that might not require us seeing the previous sol's results. Things like doing a panorama or local "remote" sensing were useful. Sometimes it might even be three sols. Or on special holidays when much of the team was elsewhere, including the engineering team, then the preplanned sols might even be longer, like a week.

The time difference restriction is not the only reason for multi-sol planning. There is one time about every two years or once every Mars year that there can be no interaction with any spacecraft at Mars, including rovers. That time is "Martian solar conjunction," that special orbital arrangement when Mars is on the opposite side of the Sun, literally. In other words, the Sun gets between Earth and Mars. Because radio signals don't travel through the Sun, when Mars is behind the Sun, there can be no communication, and the rover is literally on its own.

The time period when this takes place is about a week, but then another week or so is added on to the official or spacecraft solar conjunction to account for the fact that the radio signals get messed up or corrupted when they have to travel close to the Sun's surface. If the signals were not clear when they arrived at the rover, or if the signals were slightly scrambled, they could result in a command for the rover to do something stupid that would terminate the mission real quick. So, for practical purposes, there is about a two-week to twenty-day moratorium on sending commands to a spacecraft or rover at Mars during solar conjunction.

The constantly changing difference between Earth and Mars time is bad enough, but then there is another problem that must be accommodated. There is a problem with the words we use for time, like "day," "week," and "month." There is no "month" or even a time interval called a "week" in the Mars lexicon. These terms reflect Earth constructs that have mostly to do with the cycle of phases of the Moon and its orbit. Besides, a Mars year, that is, the time it takes to orbit the Sun once, is almost twice that of Earth, or 669 rotations of Mars on its axis as opposed to Earth's 365 rotations per year. So what are you going to do? Have twenty-four months? Or have twelve months but just make each month sixty Mars days long? And what would you call Martian months? Would a week on Mars be fourteen days long, then? Or are you going to have eight weeks per Mars month instead of the roughly four weeks of Earth's time scale? Does it all even add up to the time necessary to evenly divide up the Mars days in one Mars year, or do you need to do some sort of leap year to catch things up the way we do with an extra day in February on Earth every four years?

And then there is the word "day." A day is a significantly different interval of time on Mars from a day on Earth. As we saw before, the additional thirty-nine and a half minutes each day of a Mars day add up very quickly. And when you are operating a spacecraft on Mars it quickly becomes obvious that you need to distinguish between an Earth day and a Mars day. In the 1970s, during the first mission to successfully land on Mars, the Viking mission, it was decided to give a Mars day a different name, "sol," which, of course, is the Spanish word for sun and plays on terms such as solar, to denote one solar day. So, we can now distinguish when we are referring to an Earth day and a Mars sol. But it turns out that our language has all sorts of terms wrapped up with the concept of a "day." Words like "today" or "yesterday" are examples that we use every day. All of this was academic, of course, until a group of people, us, on the first long-lived Mars rover mission had to live and work on Mars time. We had to get creative. New words were made up on the fly, like "tosol" for "today" and "yestersol" for "yesterday." I do not think there was ever a term dreamed up for "tomorrow," except a tentatively used "nextsol"; "tosol" was already taken, and there is no Martian equivalent to "'morrow." In any case, these are the types of things you must worry about when you are pioneering a new world.

THE TAO OF TAU: MARTIAN WEATHER AND SEASONS

LIFE ON MARS IS ALL ABOUT the dust. The dust circumscribes your world and it paints your landscape. Dust will be one of the banes of human exploration. It certainly has been that with robotic exploration. This is especially true when you are a solar-powered rover and dust is wont to accumulate on your solar panels, as it does on everything else on Mars such as rocks. It was the dust that was supposed to kill us early in the mission, and did end Opportunity's journey. Previous landers had given us enough information to predict that the steady accumulation of dust would eventually blanket the solar panels of Spirit and Opportunity. And with less energy going into the batteries, eventually there would be insufficient battery charge to keep the overnight heaters going during the brutally cold Martian nights. This need for energy to power heaters was only exacerbated by the approach of Martian winter.

We began reporting the daily measurements of dust in the atmosphere and dust on the panels during our downlink reports for each sol throughout the mission. The way we measured how dusty the atmosphere was and, more important, how much the solar input to our solar panels might be was summarized in a number called tau. As part of my long-term planner duties, each day I would plot the recent trends in both and insert those in my presentation during the SOWG uplink meetings.

Tau was simply a number that stated how clear the air was. The name came from a simple algebraic equation from the atmospheric sciences that used the Greek letter "τ" (tau) to specify a comparison between an observed intensity of the Sun as seen through special filters on the Pancam and the predicted value of the intensity for a perfectly clear sky. If tau was a high number, the sky was very dusty, and if tau was a low number, the sky was clear. The number went up during the warmer seasons when things like spring winds, dust storms, and local dust swirls or dust devils would kick up enough dust to dirty the atmosphere. The number went down in the colder seasons when the convection in the atmosphere became less.

The dust on Mars is a particularly insidious dust. It is extremely fine and is more akin to smoke particles than the gritty dust many of us are familiar with in dry places here on Earth. Anyone who has lived in a region anywhere near the many fires that have plagued the world over the past few years is familiar with what a dust storm must look like on Mars—a general haze and sometimes even quaffs of smoke descending from fires hundreds of miles away. And so it appears on Mars, except that the "smoke"

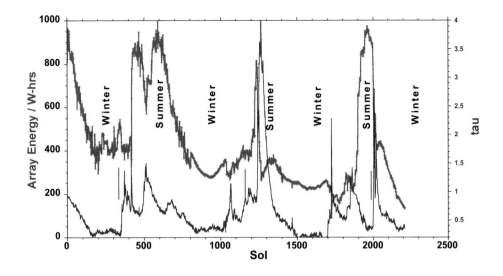

ABOVE: Plot showing the energy output from the solar panels and the tau, or dust in the atmosphere, over the first 2,209 sols for Spirit.

is dusty brown instead of bluish. And when you see Martian dust, you know it is going to settle out and coat something with a thin veil of reddish-brown. It coats rocks and solar panels equally alike.

In general, during the colder months, the winds and solar heating of the ground that cause dust-lifting events abated, the dust in the air settled out over a long period of time, and the rate at which it settled on the solar panels was greatly reduced. But during the summer months, the dust would fill the air and what goes up eventually comes down on solar panels. So we tracked the dust on the solar panels with yet another daily-reported number. That number simply compared what the panels should be putting out with the known intensity of the sunlight at Mars with what they were actually putting out in power. The assumption was that the difference was due to dust coating the panel surfaces. This was called the dust factor, and when it was 1.0, the panels were perfectly clean, and when it was 0.5, the panels were veiled with a coating of dust that allowed only half the light intensity to get through to the panels.

This number went up when we had a panel-cleaning event from some chance gust of wind. In fact, you could see the difference in rover "selfies" looking down on the panels. But for long months, during the less windy season or after a big dust storm, the number would keep going down as the panels became dustier and dustier. Watching the plot of dust factor from sol to sol was like watching your gas gauge. It was a great feeling to see that dust factor riding high, but when it dropped below 0.5 and winter was rapidly approaching you would be getting a bit nervous like driving that desolate back road with the gas gauge hovering closer to empty. It was always a happy day when we would wake up and the dust factor had suddenly jumped up. Life was good then. Then we would start watching the dust factor begin its slow descent and hope for another miraculous cleaning, if we were lucky.

THE ALIEN SKIES OF MARS

WHEN YOU LOOK AT ONE OF the typical panoramas taken by a rover on Mars there's something else other than the alien landscape that tugs at the periphery of your consciousness as something familiar. It is the lighting. Things look brightly illuminated, but the glare is not quite like that on Earth or the Moon, it is somewhat subdued. It is an almost autumnal light when the world acquires a gentle glow of warm tones unlike the burning white intensity of the previous months of summer. For this reason, I like to think that Mars is a planet of perpetual autumn lighting. On the

A typical blue Martian sunset taken by Spirit as it climbed the slopes of the Columbia Hills on sol 489 near Larry's Lookout. This view is southwest, and the Sun is setting behind the southern rim of Gusev Crater eighty kilometers away.

other hand the edges of shadows are somewhat more crisp and distinct. There may be a couple of reasons for this that have to do with the distance of Mars from the Sun and the oddly colored skies.

First, the diameter of the Sun as seen from Mars is about two-thirds of that as seen from Earth, or about twenty arc seconds because Mars is 1.5 times farther out from the Sun. As a result, the disk of the Sun is not only a bit smaller, it is, by Earth standards, closer to being like a point source than the wide disk of the Sun as seen from Earth. So perhaps shadows are a bit more focused like the shadows from a welding arc light, not quite so much, but just subtly enough to give the impression that they are different from our terrestrial kind. Second, because light intensity drops with the square of distance, the Sun appears about 2.3 times dimmer than from Earth, or about half as bright. Again, maybe the only way our human brains can assimilate this difference is by comparison with our experience in autumn when the Sun is generally lower in the sky, shadows are a bit longer, and the light is warmer in tone and richer in reds compared with the white noonday Sun of summer.

Martian sunrises and sunsets are peculiar as well. On Earth we are accustomed to the reddish color of twilight and sunrises and sunsets. Sunrises give way to the crazy blue sky of day, blue being such an unusual color in the natural world and surely a galactic wonder in itself. But on Mars things are the other way around. Sunrises and sunsets are bluish while the daytime skies are red or at least a kind of brownish.

MARTIAN NIGHT SKY: STRANGE THINGS THAT STREAK IN THE NIGHT SKY

THERE ARE STRANGE THINGS TO SEE in the Martian night sky. Over the years, both Spirit and Opportunity woke up late in the night several times to make observations of the sky, particularly passages of the Martian moons Deimos and Phobos. But other observations were made to capture images of the constellations and comets. Probably the first depiction of what it would be like to look at the Martian night sky was in the movie *Robinson Crusoe on Mars*. The idea that there may be satellites passing overhead and other wonders reminiscent of *Robinson Crusoe on Mars* immediately came to mind when we first began doing nighttime observations on Mars.

One image acquired early in the mission, on Spirit's sol 63, around 4:40 a.m. local time revealed a strange streak in the sky. I recall that there was some initial buzz among the team that it could have been a satellite captured in the fifteen-second exposure, and it was even speculated that it could be the old Viking 1 Orbiter. Later it was concluded that the orientation was probably not what would be expected given the orbit of Viking 1. So in the end it was concluded that it was probably a meteor, one of the first and certainly not the only one to be detected. Another meteor was imaged on Spirit's sol 643 in an actual effort to do so. In this case it was predicted that Mars would pass through a meteor stream associated with comet P/2001R1 LONEOS. The image consists of several superimposed sixty-second exposures separated by about ten seconds, so the star trails are dashed lines.

"Elsewhere the sky is the roof of the world; but here the earth was the floor of the sky."

—WILLA CATHER, *Death Comes for the Archbishop*

Another nighttime observing session occurred during the time when Spirit was situated on the summit of the Columbia Hills. These observations were done using the Pancam high-resolution camera. As Jim Bell was the principal investigator of that instrument and the principal advocate for these nighttime imaging sessions, we referred to them as Jim Bell's Hilltop Observatory. We captured all sorts of things, but the chief object of the sessions was Phobos and Deimos.

On Spirit's sol 594 one particular image captured the faster-moving Phobos crossing the night sky near Deimos using a series of frames spaced 150 seconds apart. A whole series of these nighttime images of Phobos and Deimos was acquired, and while they were lovely images showing what it would look like if you were outside at night on Mars the purpose was actually scientific. While we knew the orbits of Phobos and Deimos fairly well, these types of images for which we had precise timing information could increase our knowledge of the orbits considerably. This was not the only time that the satellites were observed. Successful attempts were made to capture Phobos as it eclipsed the sun on occasion as well. Later the Curiosity rover would go on to do these sorts of observations at even higher resolution. But in addition to solar eclipses at least one series of sixteen images spaced ten seconds apart was acquired on Spirit's sol 675 showing a "Phobos eclipse" as it passed into the shadow of Mars.

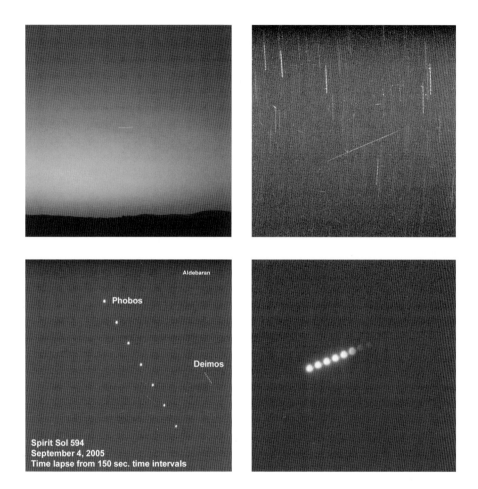

Aldebaran

• Phobos

Deimos

Spirit Sol 594
September 4, 2005
Time lapse from 150 sec. time intervals

TOP LEFT: Early on the morning of sol 63 Spirit was imaging the morning predawn sky during its attempt to capture an image of Earth when it captured this strange streak. It was initially considered that it might be a satellite, perhaps even one of the old Viking Orbiters, but later we concluded that it was a likely meteor. **TOP RIGHT:** A composite of nine sixty-second exposures by Spirit during the night of sol 643 that captured another meteor. **BOTTOM LEFT:** Sequence of images taken 150 seconds apart from Spirit's "hilltop observatory" in the Columbia Hills on sol 594, showing the progression of the moons Phobos and Deimos across the Martian night sky. **BOTTOM RIGHT:** Spirit was quite the amateur astronomer and captured this eclipse of Phobos on the night of sol 675 as it passed into the shadow of Mars in the night sky. The images were taken about ten seconds apart as Phobos moved left to right across the sky.

ELEVEN

The Invasion of Mars, Phase 3

"Knowledge is of two kinds. We know a subject ourselves, or we know where we can find information upon it."

—SAMUEL JOHNSON, James Boswell's *The Life of Samuel Johnson*

IF WE HAVE LEARNED ANYTHING COLLECTIVELY as a civilization over the past few thousand years, it is the simple fact that it is often true that to understand something well it is necessary to ask the "right" questions. And in order to get to that point, when you know nothing initially, you must simply explore. By the end of 2010 the orbital missions of the first half of the decade, together with the rovers on the surface, had explored the principal details of Mars's geologic history. This resulted in a first-order global understanding of the current planet. We now had a strong knowledge base about Mars and better understood the questions that we needed to ask than at any point previously in the exploration of Mars.

And so, we sent the next wave of missions in search of those answers. The time period after 2010 has been described as the decade of Mars because there were multiple new missions launched by several nations during the period. Some were efforts by previously successful nations to build on their past experience and to address those new questions, and others were part of a new wave of expanded international participation in Mars exploration.

Russia Phobos-Grunt ("Phobos-Ground"). This mission launched on November 8, 2011, but never left the parking orbit and eventually burned up in Earth's atmosphere. This was unfortunate because the mission's ambitious goal was an attempt at a sample return from the Mars moon Phobos. Meanwhile, the main spacecraft would continue to orbit Mars and provide new observations. There were efforts to launch a replacement later in the decade, but those efforts failed to come together. Ultimately some of the instruments originally intended for the second try were eventually flown on the ESA ExoMars Trace Gas Orbiter in 2016. But the Phobos sample-return plans were not part of the instruments.

US Mars Science Laboratory (MSL) Curiosity Rover. Launched on Curiosity on November 26, 2011, and the arrival on August 6, 2012, was a much-viewed event. There are certain missions like the Viking missions, the MER mission, the Odyssey mission, and the Mars Reconnaissance Orbiter mission that are so productive, so long-lived, and blessed with such amazing observational results and experiences that they are difficult to summarize. The Mars Science Laboratory (MSL) Curiosity mission is one of them.

Curiosity was ambitious from the start, its mission being the next step in NASA's quest for an understanding of the potential for past life on Mars. In this case Curiosity would arrive on Mars with a rover outfitted with a complex array of instruments dedicated to the search for organic compounds potentially related to past habitability. The rover itself was in the American tradition of ever bigger, as was the case for decades with automobiles. Unlike the MER rovers that were golf-cart-size vehicles, Curiosity was bigger, about the size of a small car. In fact the scale was so large that many of us were starting to refer to it not as Mars Science Laboratory—its official project name—but as Mars Science Juggernaut, with visions of a giant rover with enormous wheels rolling over the surface of Mars crushing mere rocks in its path. In reality, the rocks proved almost too much for the wheels. Early in the mission the wheels were noticed to have many punctures from the small sharp and pointy rocks. Apparently the wheel materials were a bit under-designed in the face of more rocks and sharper rocks than were originally considered. And there was also the fact that the rover landed vigorously on its wheels, inviting all the little sharp rocks exposed by the landing rocket blasts to poke forcibly into the wheels. Nonetheless there are always ways of mitigating such unexpected problems, and the rover simply takes a path that avoids unnecessary small rocks whenever possible.

The naming process followed previous naming opportunities, with suggestions being submitted by students from all over and a final name being selected from the submissions. It was Clara Ma, a twelve-year-old girl from Kansas, who gave Curiosity its name, explaining in her submission essay that "curiosity is the passion that drives us through our everyday lives." Thus the name Curiosity was selected, following the growing tradition of names that were attributes as opposed to common names. In this case the attribute is perhaps more benign. As I have suggested, some previous names appeared to have imparted some suspicious qualities to rovers. Curiosity is a name that is difficult to associate with any bad behavior. Due to the growing list of technical details that had to be accommodated during initial design and assembly, and at extra expense—chewing away at the Mars Exploration Program's total budget at the time—Curiosity

BELOW: View of the rock punctures in Curiosity's wheel taken on its sol 1,315 by the Mars Hand Lens Imager (MAHLI) camera.

Curiosity rover selfie taken on its sol 2,553. The image comprises fifty-seven individual images taken by the Mars Hand Lens Imager (MAHLI), a camera on the end of the rover's robotic arm.

was sometimes referred to as *the* curiosity in the more snarky discussions. Other than that, there have been no obvious untoward effects of the naming.

Part of the complexity was the abundance of new and more powerful instruments designed to assess the chemical and mineralogical details of the surface and, in particular, search for organic compounds. There are more than twelve instruments of unique design, including more powerful panoramic cameras, a Microscopic Imager, a rock-sampling drill, an onboard gas analyzer, and several varieties of chemical analysis equipment. The arm on the front of the rover is used to place a variety of instruments in contact with the surface for analysis. The instruments are mounted on a rotating assembly the sheer size of which has been compared with the entire size of the previous rovers alone.

One instrument that has captured the interest of the public is ChemCam, a device that is able to analyze rocks remotely by zapping them with a small laser pulse and analyzing the spectrum of the instantaneously released plasma. Roger Wiens and colleagues at nearby Los Alamos National Laboratory developed ChemCam and fought the epic battles necessary to get any instrument on a spacecraft going to Mars. Hort Newsom, a ChemCam co-investigator from the Institute of Meteoritics in the University of New Mexico up the street, is involved in the day-to-day operation of ChemCam and will be the first to tell you that ChemCam has collected an insane amount of chemical data. It is perhaps the dream analyzer for any field geologist. There is no need to hammer on the rock, collect a sample, and get it to a lab for analysis. Instead you just "point and shoot." Or so it would seem. But the reality is that in any new technology there are learning experiences and details that must be resolved to succeed. There is a lot more to it than just pointing and shooting too.

Every instrument that has ever flown on a mission has epic battles and recoveries from near fatal failures. There was one hardware problem that suddenly arose on the ChemCam in late 2014 that illustrates a "typical" disaster of the type that spacecraft instrument designers deal with on every mission, and few people outside the mission really hear much about it. But it shows the complexity and ingenuity of mission operations very well. The laser beam that ChemCam fires at a target rock must be focused according to the distance to the rock's surface. This is somewhat comparable to using a magnifying glass to focus sunlight on a surface to ignite kindling for a campfire; until the magnifying glass is held at the correct distance, the image of the Sun is too diffused to provide the focused heat necessary to burn the kindling. On ChemCam, this focusing is regulated by another small laser somewhat akin to the type that is used, for example, by contractors for measuring distances to walls, or similar to the one used by cameras to do autofocusing. Once the distance was set with the focusing laser, then the main laser beam could be focused appropriately for that distance and the "shot" could be performed.

But the focusing laser was suddenly showing signs of failing. Without the ability to focus ChemCam's more powerful analysis laser, it would be unable to acquire useful data in the future. All these instruments have some sort of backup plan, and this particular failure was considered early in the design process and filed away for future reference. There turned out to be two ways to do the same focusing measurement with existing instrument capabilities, the actual laser shot itself and the ChemCam "context camera," known as the Remote Microscopic Imager (RMI). After a lengthy analysis it was decided that some vigorous software development could be done and uploaded to link the RMI focusing to the ChemCam laser beam focusing. The story is a bit more complex than that, as any instrument story typically is, but that is the gist of the backup work-around.

There were benefits to the new method. Besides now having the ability to focus the analysis beam at much farther distances, it allowed the RMI to act like a field-spotting telescope. Objects tens to hundreds of meters away could now be "zoomed" into focus for examining important geologic relations and outcrops at a distance. This ability alone was now like having a new instrument that can be used for many new and revealing science results not previously part of the mission plans. It is a testament to typical engineering problem-solving prowess that it happens on missions more frequently than you might otherwise guess.

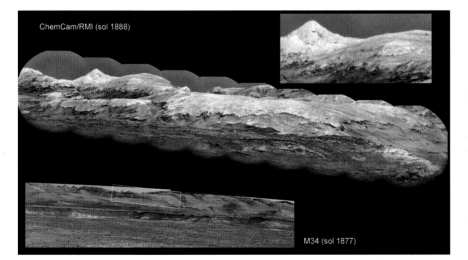

ChemCam/RMI (sol 1888)

M34 (sol 1877)

LEFT: Image of the distant slopes of Mount Sharp (Aeolis Mons) in Gale Crater using the ChemCam Remote Microscopic Imager. Mastcam image at bottom shows the normal scene from Curiosity. The inset at upper right shows the detail within a small part of the RMI mosaic.

> The history of site selection has been that evidence of sediments is evidence for preservation of things, including a historical record.

This collection of large instrument teams on Curiosity led to a return to the previous ways of doing big missions in which there is not one rover team but instead a confederation of teams each representing a separate instrument. And because of the sheer number of instruments and associated teams, along with students and associates, the entire mission team is enormous. The difference from previous rovers has been compared with a field trip of scientists in a single vehicle going from stop to stop. Whereas the MER team was like a group of scientists in a small van moving around the surface, the Curiosity rover team has been compared to a traveling caravan of several buses. Anyone who has been on a really big conference field trip knows how that works—or not, as the case may be. It can be challenging when you have a group of highly educated unlike-minded individuals examining and discussing an outcrop. The concept of the "argument on the outcrop" originated from past experiences with those situations.

Another little known fact about Curiosity is that one of the instruments that analyzed the elemental chemistry of rocks, the Alpha Particle X-ray Spectrometer (APXS), carried along a small cut disk of basaltic Earth rock as a calibration standard mounted on the front of the rover. After a wide-ranging search for a suitable standard, the rock selected was from an outcrop near Socorro, New Mexico. So we like to say that there is now a small piece of New Mexico on Mars.

Another big difference other than size and abundance of instruments from previous rovers is the fact that Curiosity is powered not by solar panels but a radioisotope thermoelectric generator (RTG), so the whole issue with dust and potentially shortened life span was bypassed with its own long-term power source. As a result Curiosity has been able to remain unaffected by things like global dust storms, or even night for that matter. One minor downside is that the radioisotope source degrades with time in a predicted way such that there will be some time several years in the future at which point the power will begin to affect its operations. In the initial weeks after landing, when all the systems and images were being checked out before commencing drives this resulted in an ironic observation. Apparently there was a growing frustration of sorts that so much time was passing before the mission got underway, and there was an alleged tongue-in-cheek comment from one engineer that the mission really should get going because "unlike Opportunity with its solar power, Curiosity does not have an unlimited supply of power." Of course, this was in reference to the phenomenal ability of solar-powered rovers to get solar energy almost forever so long as the panels kept getting cleared by cleaning events while radioactive sources of energy do decline with time. However, we all know how that played out in the end.

Given its mission objective, the choice of landing sites once again went through the community vetting process of proposal, discussion, and selection to find an area where we knew from orbital remote sensing that there was evidence for aqueous environments in the past. The history of site selection has been that evidence of sediments is evidence for preservation of things, including a historical record. And so it was with the Curiosity site, a location within a large crater, Gale Crater, where there was a stack of sediments that could be riffled through like the pages of a book by an industrious rover burgeoning with instruments designed to do so.

Gale Crater is peculiar in that the floor showed evidence for water-related minerals and right next to that, standing in the center of the crater, was a mound consisting of a stack of layers. In fact the mound was a mountain, Aeolis Mons, informally named Mount Sharp after Bob Sharp, a well-known planetary geologist from Caltech, and that mountain is actually higher than the rim of Gale Crater. The idea was to land on the floor of the Gale Crater and then rove up the stack of sediments, which likely recorded a sequence of deposition over a significant part of Martian geologic time. It was another sedimentologist's dream site.

Because Curiosity is so big and heavy, the airbag system for landing used in the past would simply not work. Instead of a rover mounted to a platform that it then rolled off and went about its tasks, in this case the rover was the lander. This new landing system garnered attention. The rover was supported in the final moments by cables attached to a rocket-powered crane, and the rover touched down on its wheels. After Curiosity had finally checked out its instruments and begun to move forward it did a look-back at the landing site. The scene was something out of an alien-landing science fiction movie: a zone with blast marks where something had clearly landed and tracks leading away. Additional drama was added by the presence of a descent-imaging camera that captured a movie of the landing process that is epic in its own right. And Mars Reconnaissance Orbiter's HiRISE camera once again captured an image from orbit of the rover as it was descending to the surface on its parachute.

While all the drama of the landing gets replayed often for Curiosity, as it has for previous missions, the mission after the fireworks of landing is the real star. It has become something of a tradition that landing sites on Mars are named in honor of someone who was a celebrated supporter of Mars science and culture. And so it was on the passing of Ray Bradbury in 2012 that the landing site was officially named the Bradbury Landing Site after the author of the classic science fiction book *The Martian Chronicles*.

TOP: View looking obliquely south at Gale Crater showing the location of Curiosity's landing ellipse on the north floor of the crater. Gale Crater is 154 kilometers in diameter and has a peculiar layered mound in its center standing five kilometers above the crater floor. **BOTTOM:** This image of Curiosity descending toward the surface of Mars on its parachute was acquired by the MRO/HiRISE camera as its orbit carried it by the landing site during Curiosity's perilous descent to the surface.

Right from the start Curiosity began making some epic observations. At one of the first sites on its traverse from the landing site it encountered an outcrop that appeared to be what we geologists refer to as a conglomerate. A conglomerate is a rock consisting of fine sediments mixed with pebbles and various rocks from different original rock outcrops that forms when water carries the material and jumbles it together as a coarse sediment. So right away Curiosity found evidence for once-flowing water that had apparently swept in debris from the surrounding rim of Gale Crater.

Later, in 2014, when Curiosity used its drill to investigate a mudstone outcrop, it acquired the first direct evidence of organic molecules on Mars. Although the molecules detected in the mudstone were likely sourced from inorganic processes, it nonetheless was a significant step in the direction of inventorying organic materials on the red planet. In this vicinity Curiosity also detected small amounts of methane. This simply added to the apparent tendency of methane detections to come and go. The jury is still out on the Mars methane mystery.

In 2015, as Curiosity was working its way to the base of Mount Sharp, John Grotzinger, from Caltech, the Curiosity project scientist at that time and a leading authority on sedimentary processes on Mars, and his colleagues published a short paper in *Science* outlining the observational goals of Curiosity's unique sedimentary setting. Orbital remote sensing with instruments like MRO's CRISM had mapped an interesting mineral abundance sequence in a series of layers near the base of Mount Sharp. The layers recorded a history of changing environments laid down by materials washed in from the rim of Gale Crater and previous intermittent episodes of floor-covering lakes. The interesting thing about the sedimentary layers as seen from orbital remote sensing was the presence of a sequence of mineral compositions that recorded the changing global climate and chemical environments of Mars. Near the base were clay-bearing materials and near the top of the sequence were sulfates, very much like the Phyllosian, Theiikian, and Siderikan geologic time sequence discussed in Chapter 5. It was a physical record accessible in one place of the mineral eon sequence of Martian global geologic stratigraphy. Visiting and studying such a sequence would provide the all-important test of the proposed changes in global geologic environments over time. Testing hypotheses is what science is all about, and this sequence of sediments offered a test of an important hypothesis about Mars's geologic history. As of Spring 2021, Curiosity has been winding its way up the sedimentary stack near the northwest base of Mount Sharp, systematically investigating each unit.

India Mars Orbiter Mission (MOM), or Mangalyaan (Hindi for "Mars Craft").
Launched November 5 and arrived on September 24, 2014, this mission is a remarkable success story and its Mars Color Camera (MCC) has acquired some spectacular images of Mars. Although it was Indian Space Research Organization's (ISRO) first interplanetary space mission and is described as a technology demonstration rather than an attempt to do new in-depth science, they nailed it on the first try, a thing that no other nation has been able to achieve. Not only that, it is the least expensive Mars mission to date, coming in at $73 million. This is practically the cost of the launch rocket alone for many space missions initiated elsewhere.

The science component of the mission is the study of the geology and surface morphology of the red planet as well as making observations of the atmosphere, including methane, and exploration of the effects of the solar wind and atmospheric escape on the upper atmosphere. Many of the Mars Color Camera images are suitable for framing or as a cover image for a book about Mars they are so lovely.

Mangalyaan has been a "happy mission" too, a rare thing for a Mars mission based on our collective experience with all the failed attempts over the years by several nations. From the outset there have been no major glitches or hardware "events" and the spacecraft has done everything that the controllers asked of it. This is an amazing thing in itself, but on a first mission it is outrageous. One can only guess what Coyote Mars has in store for this one. Perhaps the object is to give the Indian Space Research Organization a "pass" on this first effort in the hopes of luring them into the next step or "learning experience," whatever that might be.

US MAVEN (Mars Atmosphere and Volatile Evolution). MAVEN is a detective mission, starting on November 18, 2013. It was not an imaging mission concerned with the current surface geology of Mars like many previous missions. Instead it was designed to address questions about the Martian atmosphere left unresolved by those previous missions. Its goal was to find out what the heck had happened to Mars's atmosphere. Bruce Jakosky, Laboratory for Atmospheric and Space Physics, University of Colorado, the principal investigator, is a quiet sort who steadily plugs away and had been trying to kick-start the atmospheric end of Mars exploration for years. The time was right, and he got the mission on the program.

Evidence from many previous missions had shown us that Mars once had abundant water and perhaps a thick atmosphere that would allow water to exist on the surface. But water and gases are stealth materials; they do their thing and then vanish from the scene. The only way to determine if water was there, how long it was there, and what happened to it is to do some clever investigative work. In this case the goal is to find out what happened to the atmosphere and how. The answer would support a better understanding of Mars global history and help resolve many outstanding mysteries about the past climate. Yet to date no mission had the capability of sifting through the clues in the remaining atmosphere. Fortunately the atmosphere today holds some clues in its chemistry. MAVEN's goal was to do just that by orbiting Mars with the instruments designed to search for chemical clues in the modern atmosphere.

The MAVEN mission is currently continuing to gather data, but after the first year, initial results suggest that the atmosphere was lost through erosion by solar radiation, a result of the absence of a protective magnetic field like we have on Earth. More to the point, a significant amount of this loss happened in the first five hundred million years, which accounts for the rapid decline in the surface evidence for water early in Mars geologic history.

ESA ExoMars Trace Gas Orbiter (TGO) / Schiaparelli Rover. The latest Mars mission from the European Space Agency included both an orbiter and a lander. The mission launched on March 14, 2016, and eventually arrived in February 2017, followed by a lengthy process of trimming the orbit to a circular working orbit using the process of aerobraking. The method of aerobraking had been used by previous spacecraft and relies on dipping into the upper reaches of the planet's atmosphere on first approach to the planet or on multiple orbits to gradually change the orbit by slowing the spacecraft down. The final orbit of nearly four hundred kilometers altitude for TGO was attained with some additional thruster firings.

LEFT: A visualization of the extent of methane detection on Mars.

RIGHT: An image of methane concentrations in North America via a remote sensing satellite. The highest concentration is in an area of gas production in northwest New Mexico.

Methane release:
Northern summer

Methane Concentration

0   5   10   15   20   25   30
parts per billion

The primary goal of the orbiter was to study the gases in the Martian atmosphere that may be evidence for extant life. Methane (CH_4) had been detected several years earlier in 2004 by the ESA Mars Express spacecraft and subsequently by several Earth-based telescopic measurements. Methane is regularly monitored in the Earth's atmosphere in much the same way.

The observed presence of methane is interesting because it goes away quickly in the Martian atmosphere, and its presence requires that it be generated today. And because it is not always detected, there is the suspicion that it is generated sporadically, like there is an active source. But where and how is it generated?

Most of us are familiar with the fact that methane is a biologically generated gas. Cows for example are a familiar source, but so are certain microorganisms. So the detection of methane at Mars suggests the possibility that there is an active microbial community somewhere. But methane is also a well-known product of certain nonbiological processes. Methane is present in some volcanic gases and another big source includes the process of alteration of certain iron- and magnesium-bearing rocks like those that make up the upper mantle. The process is called serpentinization and involves deep water and carbon dioxide interacting with minerals such as olivine that are major components of deep crustal and mantle rocks to form the mineral serpentinite. And, of course, olivine, as one of the major minerals in basalt, is an important mineral on the surface of Mars as well as its mantle. Without additional information we cannot determine whether the methane is biological or nonbiological in origin.

Because of the whole question of methane in Mars's atmosphere, TGO's principal goal then was to determine if the detections were real and if so when the release events occurred, and where they were being released on Mars. With the exception of a possible detection, TGO has not at this point made a significant methane detection. While Curiosity did detect methane on a couple of occasions, TGO saw nothing even though it was observing the surface in the vicinity at the time. Methane is proving elusive. Either the methane is sporadic and local, or something else is going on.

The Schiaparelli rover that was part of the Trace Gas Orbiter mission was released from the orbiter and attempted a landing on

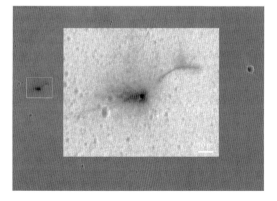

October 19, 2016. The site chosen was Meridiani Planum, not far from the site where Opportunity had landed in 2004. Schiaparelli was a technology demonstration, and the Meridiani Planum site was selected because it was a known safe location. On the day of the entry, descent, and landing we were planning to see if Opportunity could see the entry somewhere in the western sky. It would be the first attempt by a landed rover to capture the landing of another spacecraft on Mars. But unfortunately Opportunity was just below the rim of Endeavour Crater at the time and that part of the sky was blocked from view by the rim. In the end there was a problem that prevented the rockets from slowing the lander/rover down and it crashed.

US InSight (Interior Exploration using Seismic Investigations, Geodesy, and Heat Transport). I have run into Bruce Banerdt, Jet Propulsion Laboratory, the principal investigator of InSight, on many occasions over the years. A fellow traveler within the lost generation, he reminded me of Bruce Jakosky, the principal investigator of MAVEN, in some ways. He is confident of the geophysical science that needs to be done on Mars but frustrated by the absence of a seismometer on Mars and felt that it left a gaping hole in our understanding of the planet. We have entered an era when some of the gaps in our knowledge of Mars are starting to get in the way of progress with understanding Mars as a whole. One of the big questions is the interior of Mars, and a good way to get at that information is the same way we do it on Earth, through the observation of earthquakes, or in the case of Mars, marsquakes.

Another missing element in our study of Mars is the interior heat flow. Internal heat of a planet is why things get hotter as you go deeper within the planet, and it is a fundamental property that says something about the energy available for both earthquakes and volcanism among other things. Without a measurement of the heat flow from the interior of Mars we are left guessing about the current state of the planet. Thus the InSight mission came about to resolve the seismicity and heat flow of Mars.

The landing site for InSight was chosen with the requirement that the site be quite bland and free of rocks. This was not a surface geology mission, and the desire was to land in a site where there would be a terrain that the seismometer could be deployed. Just as important, the site requirements included the need for sufficient soils such that the Heat Flow and Physical Properties Package (HP^3) could operate. We have become very sophisticated in the landing site certification process over the years, and once again Matt Golombek and colleagues got to work sorting through the candidates. The analysis of landing sites has gotten so detailed, and uses data from so many different sources, that the certification documents challenge the ability of most

LEFT: InSight lander selfie taken on its sol 10 after landing on Mars.

BELOW TOP: Part of a panorama acquired on sol 10 of the surface at the InSight site on Mars. The surface was selected specifically for low rock abundance and bland terrain.

BELOW BOTTOM: Image of InSight's seismic station deployed on the surface of Mars.

mortals to follow analyses in that process. I was exhausted just reading the final landing site certification documents. It was a kind of overkill, in a good way, with each data set being hammered to a pulp and then the next one similarly hammered. And this went on and on. About the only thing left to do was go to the finalist on the site search and confirm the predictions. There was little room for the predictions to be very far from reality. In the end the site chosen was not too far north of where Curiosity landed, but in this case a fairly old surface of Elysium Planitia north of the highlands bordering the Utopia basin. It is a geologically rather unexciting place, but perfect for a lander that requires simple terrain.

InSight successfully landed on November 26, 2018. The lander was similar to the Phoenix lander, using parachutes and final descent rockets to softly touch down on the surface. An image panorama revealed that the site was as predicted and that InSight had landed in a small hollow where there were few rocks and plenty of soil for deployment of both the seismometer and the heat flow probe.

Once the seismometer was successfully deployed to the surface it began its long-awaited quest for marsquakes. Several events have been detected, including those related to surface weather such as dust devils and thermal cycling of components of the spacecraft. But many clear marsquakes have been detected. Results from the first

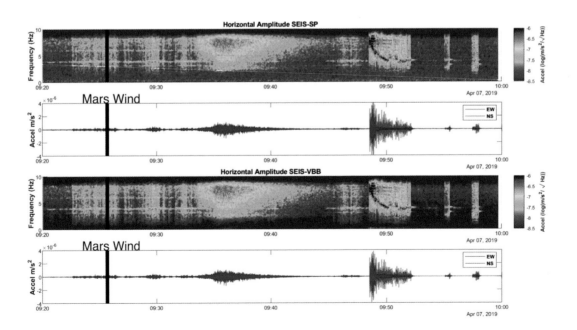

year of operation were reported in February 2020. The largest quake observed was estimated to be a magnitude 4.0 and appears to have come from the region of Cerberus Fossae in Elysium Planitia. Cerberus Fossae is a very young deep chasm-like fracture feature in the area south of the big Elysium volcanoes, Elysium Mons and Hecates Tholus. It is the same area where some of the youngest lava flows are known to have occurred on Mars, some as young as ten million years. So it is not surprising that we should detect marsquakes from this region. Once a suitable period of time has elapsed, and hopefully a much larger quake is detected, all these observations have resulted in a new model for Mars's interior structure. A large quake is necessary to really get a good look at the deep interior.

The heat flow probe has had a bit more difficulty getting started. It was supposed to self-hammer, like a pile driver, using a "mole" device. The plan was to go down to five meters below the surface and out of the influence of daily temperature events, where it would detect the heat from the interior. But the mole had considerable difficulty getting started. Apparently the soils are slightly tougher than that encountered at previous landing sites such that the mole was making very little progress. Eventually, through judicious use of the lander's arm pressing down on top of the mole, it finally penetrated the soils somewhat and was fully buried. Unfortunately, the mole failed to penetrate any deeper into the local soils of Mars and in early 2021 that experiment was terminated. Heat flow measurements must await a future mission.

LEFT: Seismic signal from InSight of a possible marsquake detected on its sol 128. A plot of wind at the same time is shown beneath the two seismic measurements in order to subtract from the total signal any disturbances in the surface due to wind. **ABOVE:** The silver-colored heat probe or "mole" as seen on InSight's sol 333 as the robotic arm was being used to assist the mole in penetrating into the surface.

TWELVE

Future Mars: Mars Exploration Next

"The last sound on the worthless earth will be
two human beings trying to launch a homemade space ship
and already quarreling about where they are going next."
—WILLIAM FAULKNER

AMBITIONS FOR THE EXPLORATION OF MARS have been kindled more and more by an explosion of astrobiology research. The extent of new astrobiology research and publications is akin to the growth of planetary volcanology research in the early decades of the space age. While the desire to send humans to Mars remains a top item in the bulleted list of Mars plans, albeit an item with no firm start-up plans for the moment, there have been efforts from several new nations to engage in the popular human pastime of exploring Mars. The current decade promises to be an exciting one in that regard. Humans are expanding their presence in the solar system, and Mars is becoming abuzz with a swarm of spacecraft.

THE SUMMER OF MARS

THE YEAR 2020 WAS ONE OF those opportunities that occurs every twenty-six months for launching missions to Mars. In addition to the United States' much-anticipated Mars 2020 mission, known as the Perseverance rover, two other nations new to the Mars armada have entered the grand quest. The seemingly relentless series of new launches throughout 2020 was frequently referred to as the summer of Mars by the media, culminating with three orbital insertions in February 2021, a virtual invasion fleet.

UAE Hope (Emirates Mars Mission Orbiter), launched in 2020, is the first interplanetary mission from the Emirates Space Agency and has many purposes. The name was chosen for a similar reason as was the name Nozomi (Japanese for "hope") for an earlier Japanese Mars attempt. It was chosen simply because the mission itself represents a message to everyone in the Arab world, a message of hope about their future. The mission itself will orbit Mars with the science goal of studying the atmosphere, but it serves many other purposes beyond that goal. In addition to inspiring future Arab generations, it is "hoped" that it will build the technological capability of the Emirati in space exploration in general, foster the future of international cooperation in Mars exploration, and in general build the Emirates' position of technological authority in the Arab world. Much of the science exploited by the Western world

LEFT: Image taken by UAE Hope on arrival at Mars. RIGHT: Artist's publicity image of the Tianwen-1 orbiter and the Zhurong lander/rover at Mars.

over the past few centuries actually originated in the Arabic world. So rather than an attempt to "catch up" with the rest of the technological world, the Emirates Space Agency mission to Mars simply represents the Arab world's reclaiming of their position of authority in science and technology.

Although the mission goal of studying the atmosphere may appear modest in comparison to some of the exotic hardware that has been thrown at Mars over the last two decades, there is a purpose. To achieve those broader goals the Emirates Space Agency wanted to start with a realistic mission, and one that had a solid chance at success. Nonetheless the spacecraft will deploy a number of new investigations not previously attempted in a quest to follow the daily and seasonal weather cycles on Mars.

The purpose will be to establish the link between weather in the lower atmosphere and processes that lose hydrogen and oxygen in the upper atmosphere. This follows from the continued desire to understand how Mars started out with so much water and what appears to have been a thicker atmosphere but gradually transitioned over billions of years to the dry and thin atmosphere of the present. The spacecraft will orbit for a period of fifty-five hours and a low inclination to the Martian equator, which is an unusually low inclination that will allow the instruments to observe the atmosphere over different times of the Martian day. Instruments include two spectrometers, one in the infrared and another in the ultraviolet, and one imager in the visible and ultraviolet. Together with the US MAVEN orbiter, Hope should build our knowledge of a part of the Martian story that has been missing.

China Tianwen-1 is an ambitious mission that sent an orbiter, a lander, and a rover, Zhurong, to the surface of Mars. Tianwen-1 ("questioning the heavens") is China's first Mars mission but is part of a long-term goal to build the technology and the science necessary for future Mars missions by China to the red planet. While China has now successfully landed on the Moon with rovers, this mission is particularly challenging because landing on Mars requires heat shields and supersonic parachutes that add additional complexity to the process. Previously only the United States had successfully landed and operated anything on the red planet.

Zhurong will land in the area of southwestern Utopia Planitia, about eighteen hundred kilometers southwest of the site of the 1976 Viking 2 Lander and about six hundred kilometers from the southern highlands. This is an area of extensive plains with a variety of evidence for unusual, perhaps past ice-related processes. The solar-powered rover includes a ground-penetrating radar that will search for evidence of shallow water ice. Also on board are cameras and remote chemical analysis instruments somewhat similar to the ChemCam of the US Curiosity rover. Meanwhile the orbiting spacecraft carries high-resolution cameras, an orbital radar mapper, and other instruments that will provide additional new information about the Martian surface.

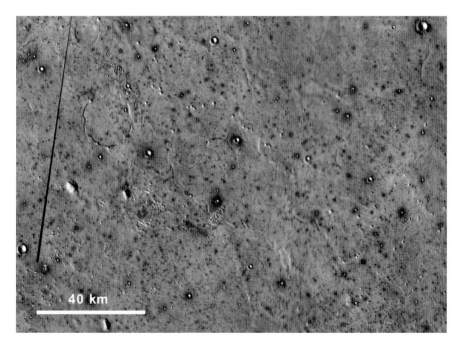

ABOVE: Odyssey/THEMIS image of the proposed Tianwen-1 landing site in southwestern Utopia Planitia.

U.S. Perseverance (rover/Mars 2020) arrived on February 18, 2021 after a July 2020 launch to Mars, the preparation for which required surmounting enumerable obstacles associated with vehicle delivery during a global pandemic. It culminates NASA's step-by-step process of searching for signs of past habitable environments on Mars and takes it the final step. It is designed to explore an ancient river delta in Jezero Crater on the northwestern rim of Isidis Planitia where it will explore the sediments within that delta for signs of past microbial life using several powerful instruments capable of detecting organic compounds and imaging rocks in extreme resolutions. It might be described as a great fossil hunting field trip to Mars, except in this case the fossils would be microbial rather than shells and bones.

Perseverance builds on the design used by the previous Curiosity rover and adds to it several new and even more powerful instruments as well as new rover driving capabilities that facilitate even more rapid traverses of the surface. The principal objectives of the mission are to search for signs of past habitability in the rocks, actually seek evidence for "biosignatures," cache well-documented samples of important visited sites for a possible future sample return mission, and test oxygen production methods in preparation for potential future human missions that would benefit from in situ resource utilization. The name "Perseverance" was the winning name proposal from over 28,000 submitted in a nationwide naming contest. A seventh-grade student from Virginia, Alexander Mather from Lake Braddock Secondary School in Burke, Virginia, submitted the name in his essay.

In addition to the instrument array needed for characterizing the surface and weather, Perseverance carries several additional technology demonstration experiments. These other technologies are designed to support intelligent obstacle avoidance during landing and the addition of a small, drone-like helicopter, named "Ingenuity" to explore the potential of aerial vehicle in future Mars missions for advanced mapping out of rover traverse pathways. The name "Ingenuity" was submitted as part of another naming contest by Vaneeza Rupani, an 11th grader at Tuscaloosa County High School in Northport, Alabama.

I was fortunate to be part of a small science team that originally proposed the helicopter as an instrument capability for Perseverance. However, the proposal as an instrument was unsuccessful when other instruments were selected in a standard competitive instrument selection process in 2014. Nonetheless the idea of an aerial platform to increase the ability to explore beyond where a rover could easily go was powerful enough that it was decided to add a scaled-down version of the helicopter as a "technology demonstration." Technology demonstrations, not technically being

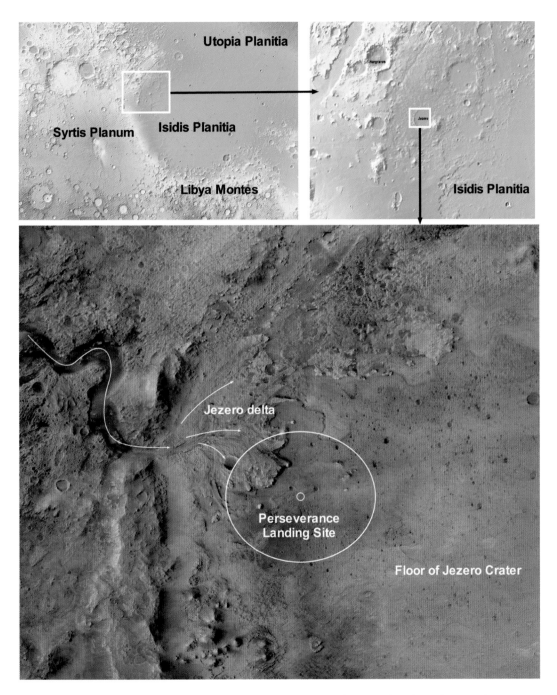

Utopia Planitia

Syrtis Planum

Isidis Planitia

Libya Montes

Hargraves

Jezero

Isidis Planitia

Jezero delta

Perseverance
Landing Site

Floor of Jezero Crater

ABOVE: The image on the upper right shows the location of the image on the upper right on the northwest margin of Isidis Planitia. The box in the upper right image shows the location of the image shown at the bottom. The landing site for the Mars 2020/Perseverance rover near an ancient delta within Jezero Crater on the northwestern margin of the Isidis basin.

ABOVE: The Mars helicopter Ingenuity in a clean room at JPL.

science instruments, do not have science teams. So that small science team was dissolved, and my future on the mission became unknown at the time. We were told that we would have to reapply for the team in what was likely to be a formidable competition for the few slots available on the final mission science team.

While the nominal mission is planned for one Mars year, being powered by an RTG like the Curiosity rover means that it has the potential to survive and explore Mars' surface for several years. During this time Perseverance will traverse across the plains and work it ways towards the foot of the delta exploring all manner of sedimentary environments. On the Earth ancient deltas are recognized by their consistent pattern of sedimentation associated with the way deltas build out from the mouth of the river channel into an original lake or sea.

Since my original "seat" on the mission as a member of the helicopter team evaporated when the helicopter was de-scoped to a technology demonstration, the only way to get back on the mission was through a later program called "Participating Scientists." Unfortunately, as noted, that program promised to be highly competitive, and for good reasons. We had shown with the MER mission and Curiosity rover how rovers were so much fun and exciting that suddenly everyone wanted to be part of a rover mission somewhat like everyone wants to become and astronaut. This meant that the competition for the few "seats" on the mission would be significant.

Fortunately, in the end I was able to field a winning proposal for a slot on the team; another close call there. So, I came on board the Perseverance mission in 2020 as part of a group of 13 competitively selected scientists whose purpose is to add additional scientific expertise to the already existing teams associated with the mission instruments. It is my duty on the mission to act as a field mapping geologist using the rover's eyes and tools to map the geology along the traverse and help make decisions about where to go and to understand the context of important sites of outcrop campaigns. It will be yet another opportunity for me to "walk on Mars." It has been a long journey from being a kid waiting by the radio for news of the first flyby of mysterious Mars by the Mariner IV spacecraft in 1965 to another grand expedition to the surface of the red planet.

Perseverance successfully landed in Jezero Crater after the now well-known harrowing sequence of events associated with entry, descent, and landing. It was the ninth time that the U.S. has successfully landed a spacecraft on Mars. Again the Mars Reconnaissance Orbiter camera HiRISE caught the parachute and rover descending toward the surface with the delta of Jezero Crater in the background. During the landing multiple cameras on the rover itself captured the spectacular final moments

Artist depiction of the Mars 2020/Perseverance rover on the surface of Mars.

in which the rover was unreeled from the hovering sky crane and lowered to the surface amid a storm of dust kicked up by the rocket exhaust. Although spectacular, the cameras were actually part of the "Terrain Relative Navigation" system that compared the images taken during descent with on-board maps of the area in real time in order to divert the lander to the safest location for landing. The process used artificial intelligence to do pretty much what Neil Armstrong did during the last moments of landing on the Moon in 1969. At that time the otherwise automated landing process was sending the Eagle lunar lander toward a boulder field and Neil Armstrong took manual control to divert to a smoother area. In this case Perseverance did an assessment of several possible sites during the last minutes of descent and calculated the best fit on each for safety. Of the two sites identified by the automated process, that happened to be a bit farther out from the primary Jezero Crater delta target, and was in the last seconds deemed a better match by a few percent. You could almost see it doing this in the landing movie, briefly hovering over a spot between the two sites and finally selecting one and going for it. This new controlled landing process essentially enabled shrinking the landing ellipse to a ten-kilometer circle, far smaller than we had with the Viking and even the Spirit and Opportunity or Curiosity landing sites.

One of the most complex and delicate roving machines assembled by humans, weighing over a ton, had arrived safely and ready to explore in a new world.

Exploring Mars this time as a team was very different from the way we had done it in the past. Due to the continuing pandemic in early 2021, we on the science team were all attending to activities via remote meetings and flight software interactions via internet. While this has become common practice during the later stages of missions before, it has been customary in the past for the entire team to be present at the Jet Propulsion Laboratory for the early months of a new mission. In many ways the isolation made it feel even more like we were lonely explorers suddenly landing on another world far away from Earth. People were all on screens and far away with a variety of communication sessions separated by solitary activities associated with mission tasks.

Another fundamental difference is the size of the mission, not just physically, but its greater complexity and number of team members. Whereas earlier in Mars exploration we were a small band of "mountain men" exploring a new world, here we were returning to set up a large outpost with many different specialists each with a separate if parallel set of tasks. Just the architecture of the software and the array of software tools and internal communication channels is far beyond the simple one-on-one team interactions we had used on previous missions. I have a fear tucked away in my thoughts that along with the greater complexity there may be an increased risk of error or miscommunication. But perhaps that fear is that of an old mountain man tisk-tisking at the arrival of a crew to construct one of those newfangled railroads. The culture is different and the sense of isolation in the world of a pandemic was nonetheless a new experience for me in space exploration.

In the days following landing we could see the cliffs of the delta and, beyond that, the towering rim of Jezero Crater, both eventual targets of Perseverance's exploration. The immediate tasks consisted of two things. One was the collective desire to understand the landing site geology, a process complicated by the need in the initial days to check out instruments and perform a variety of other "health checks."

ABOVE: Part of the panorama acquired by Perseverance on sol 3, day 3 after the successful landing. This view is looking west toward a 200-meter-wide remnant of the sedimentary delta deposits in Jezero Crater two kilometers away and the mountainous rim of Jezero Crater fills the sky beyond. The rock in the closer middle ground is about two meters across and 130 meters away.

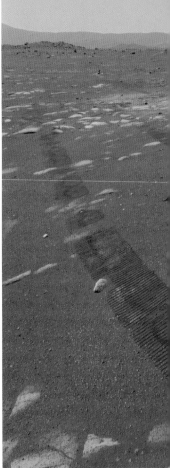

Before commencing exploration and the mission science goals, we needed to set up the helicopter experiment. To do the helicopter tests we needed to park for about 30 sols while the tests were taking place. But before we could park, we went through a series of drives over the course of the 60 sols prior to parking for the helicopter test flights, first to drop off the debris shield covering the helicopter that needed to be some fifty meters or more from the helicopter flight zone. Then we had to drive back to a location chosen for the helicopter "airfield" to place the helicopter on the ground through a delicate series of operations over several sols. This was followed by a quick "selfie" with the helicopter next to Perseverance before driving away to the east of the helicopter for the 30 sol wait for helicopter activities. With all this driving back and forth there was little time to knuckle down to actually using the array of tools we have on the rover, including touching and analyzing rocks. Although we were 60 sols into the mission and we still did not know much about the rocks, we had begun analyzing the Jezero delta from a distance and making plans for drives that would get us to it.

The Ingenuity helicopter successfully flew for the first time early on the morning of April 19, 2021, and went on to complete several additional and more ambitious flights in subsequent sols. Much of the preparation on the science team end was focused on planning the zoom camera (Mastcam-Z) pointing and timing to acquire some high-resolution movies of the epic event. Although the first flight was

ABOVE LEFT: Perseverance and Ingenuity on Mars in a "selfie" family portrait taken on sol 42 after placing Ingenuity on the surface.

ABOVE RIGHT: The Ingenuity helicopter flying on Mars on sol 61, its second flight, here hovering five meters above the surface as viewed by Perseverance high-resolution camera Mastcam-Z. The western rim of Jezero Crater forms the "mountains" in the distance.

a simple up-and-down maneuver, it was an historic first powered flight on another planet and it demonstrated that aerial flight on Mars is possible, opening up the potential for future missions in which a helicopter could be used for scouting ahead of rovers during exploration or as a platform for surveying much larger areas on Mars than a rover can traverse. In honor of the fact that we here on Earth had done the first official power flight only 117 year ago, Ingenuity carried along a small fragment of fabric from the original Wright flyer. This means that a small piece of the first powered flyer on Earth took a second trip on another first flight, but this time on Mars.

And so a new journey of exploration begins. We have substituted our coonskin hat with a pith helmet now, but it is still unknown country never before seen by any previous explorer. Who knows what wonders we will find in the hills of Jezero delta and beyond?

Mastcam-Z panorama from the landing site.

Mastcam-Z 110 mm zoom lens view of the Jezero
delta margin from the landing site on sol 4.

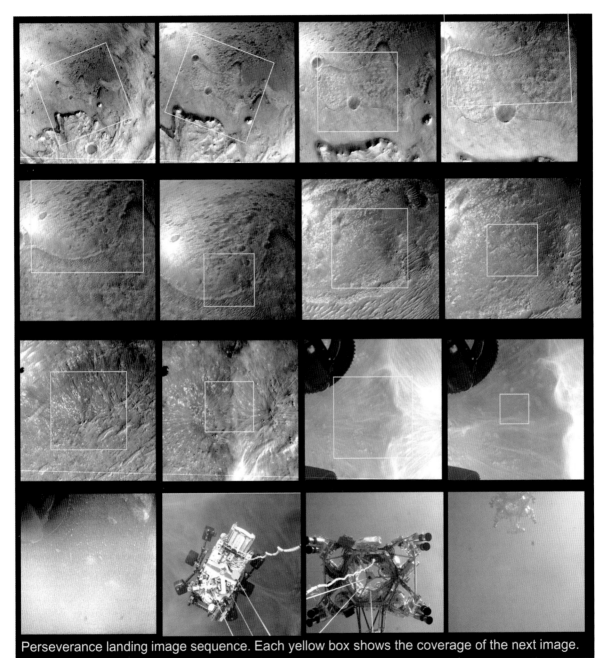

Perseverance landing image sequence. Each yellow box shows the coverage of the next image.

ABOVE: Sequence of images from the Perseverance's landing approach.

TOP: Mars Reconnaissance Orbiter HiRISE camera view of Perseverance on it parachute and its way down to Jezero Crater. **BOTTOM:** The Hazcams on Perseverance captured this view of the plume rising over the local horizon from the crash site of the landing rocket stage.

TOP LEFT: View of the parachute as Perseverance decelerated in the atmosphere during descent. The pattern of color segments spelled out a message in binary code: "Dare Mighty Things" along with the latitude and longitude coordinates of JPL. NASA/JPL-Caltech. **TOP MIDDLE:** Mastcam-Z view of the rocky escarpment east of Perseverance on its sol 62. **TOP RIGHT:** Ingenuity's downward-looking Navcam view of the ground as it flies over its "air field" on its third flight.

BOTTOM LEFT: Mastcam-Z view of the Mars helicopter "Ingenuity" awaiting its flight after being set on the ground by Perseverance. **BOTTOM MIDDLE:** Mastcam-Z view of the landform "Kodiak" several kilometers west of the landing site. Kodiak is thought to be a remnant of the Jezero delta. **BOTTOM RIGHT:** Ingenuiy's color camera looks out to the side during flight to capture a detailed view of the surface. Note rover tracks left during Perseverance's maneuvers prior to Ingenuity's flight tests.

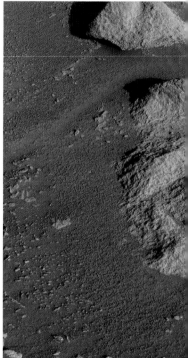

LEFT: As Ingenuity approached its second and new airfield to the south of Perseverance on its fourth flight it acquired this color image of its destination landing site where it would land on its fifth flight a few sols later.

ABOVE: Ingenuity captured this color camera image of Perseverance 85 meters away. Outlined here is a white box in the upper left and expanded on the left. Ingenuity was flying about 5 meters off the ground on its third flight. The inset shows the orientation of Perseverance as seen in the image.

RIGHT: Sequence of images from the movie of Ingenuity's fifth flight to its landing site south of Perseverance.

BELOW: Peculiar weathered–appearing outcrops in front of Perseverance on sol 66. The rocks are fractured, appear to have very granular and somewhat "crumbly" surfaces, and occur as polygons separated by fine-grained regolith. Mastcam-Z image.

LEFT: Matscam-Z image acquired on sol 68 looking north and east toward a large hill "Santa Cruz" 2.5 kilometers away.

THE NEXT GOALS
2022

ESA Rosalind Franklin rover was originally scheduled for launch in 2020, but the team encountered problems with testing of the descent parachute. As a result, the mission is now scheduled to launch in the fall of 2022. As with NASA missions ESA sourced the name Rosalind Franklin through a public-outreach campaign. The winning name was chosen to honor the English scientist Rosalind Franklin, who was a co-contributor to the understanding of the DNA molecular structure.

The Rosalind Franklin rover is actually part of a joint venture between ESA and Russia's Roscosmos, with Roscosmos supplying the lander from which the rover will be deployed. The rover is a solar-powered six-wheel design a bit larger than the US Spirit and Opportunity rovers but smaller than the US Curiosity and Perseverance rovers. Like Perseverance, a primary mission objective of the Rosalind Franklin rover will be to search for biosignatures on the surface of Mars. The site currently selected is in the Oxia Planum region on the east margin of Chryse Planitia. This area is a place where channels have formed as water once flowed from the highlands to the south into the lowlands of Chryse Planitia. Selection of the site was a result of the desire to

ABOVE: Artist view of Rosalind Franklin rover on the Kazachok lander after arrival on the surface of Mars in the Oxia Planum region. middle: Artist's rendering of the MMX Phobos rover after deployment to the surface for sample collection. **RIGHT:** Artist's concept of the MMX orbiter at Mars's moon Phobos.

land in an area of ancient sediments where not only orbital remote sensing indicated morphological evidence of past water but also spectral data indicated the presence of minerals that are associated with past water. However, due to the delay in launch the final landing site selection may change.

Russia's Roscosmos Kazachok ("little Cossack") lander will remain stationary after the rover is deployed and continue to make a variety of observations of the landing site including a seismometer to measure marsquakes.

2024

Japan Martian Moons eXploration (MMX) is a probe that will launch in 2024 with the objective of returning a sample of the Martian moon Phobos. The mission was developed by the Japanese Aerospace Exploration Agency (JAXA). In many ways it is similar in objectives to the Russian Phobos-Grunt mission of 2011, as an orbiter will be placed around Mars carrying a small rover. The orbiter will visit both the Martian moons, Deimos and Phobos, and eventually deliver a small rover to the surface of Phobos to collect loose soils and rocks, which will then return to the orbiter. The orbiter will make additional flybys of Deimos, and then the collected Phobos samples will be launched back to Earth.

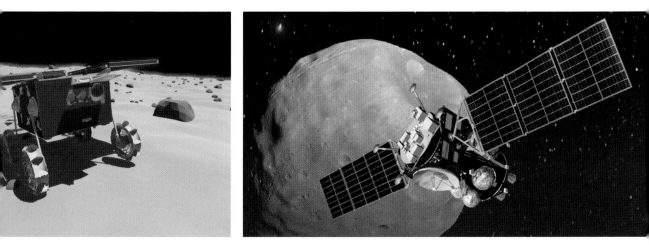

The Phobos rover/sample collector is a microwave-oven-size vehicle, or about the same size as the 1997 Mars Pathfinder rover Sojourner, and is being developed by the German and French space agencies, Deutsches Zentrum für Luft- und Raumfahrt (DLR) and Centre national d'études spatiales (CNES), respectively. It is thought that in addition to samples of the moon Phobos, it may be possible to collect small samples of the surface of Mars that may also reside on Phobos. Over billions of years it is possible that pieces of Mars's surface ejected by large impacts have been swept up by Phobos. If so, we could get a twofer, samples of both the mysterious moon Phobos and actual samples of Mars. The samples would be returned in 2029 in the time frame similar to the proposed return of Mars surface samples cached by Perseverance, as outlined in current NASA Mars Exploration Program goals.

2026++

Mars sample return, frequently referred to as MSR, has been a goal of Mars exploration for several decades. There are many reasons for a sample return, and there are just as many challenges to doing one. The return of samples to Earth would enable the use of a wide variety of more complex laboratory tools for understanding Mars, tools that simply cannot be used in a robotic in situ setting. Getting the samples requires some exotic engineering. As a result, several mission architectures have been proposed for doing so. But the most recent scenario involves a multi-mission orchestration of activities.

First, there is the collection of the samples. Under NASA's current plans the sample caching that will be part of the Perseverance rover's mission is designed to provide well-documented samples that could be retrieved with a subsequent mission and returned to Earth. This is an effort to streamline the process that would otherwise require an entirely separate mission to collect the samples, a process that would only complicate the engineering and expense. Both NASA and ESA are formulating plans for a follow-on mission that would retrieve the samples with a launch in the 2026+ time frame. The lander would be targeted for the Perseverance traverse region and retrieve either the cached samples, or the Perseverance rover would rendezvous with the lander with retained caches for placement on an ascent rocket. This would then be launched into Mars orbit and subsequently rendezvous with an orbiter potentially supplied by ESA. The orbiter would then return the samples back to Earth for retrieval either in Earth orbit or direct return to Earth.

ABOVE: Concept drawing of sample-return vehicle lifting off from the surface of Mars.

One enormous complication to the process is the desire to ensure that the samples are not in any way exposed to the Earth environment in the process. Of course, one of the principal goals for sample return is to use advanced laboratory methods to search for possible past life in Mars samples. Therefore, any contamination from Earth's environment runs the risk of spoiling the search for non-Earth life in the samples. The other more important concern is the possibility that Mars organisms, should they be present, might contaminate Earth. While that possibility is judged to be remote, as it was with the Apollo lunar samples, nonetheless such an event, however remote, is considered potentially catastrophic enough to warrant some serious efforts to prevent it from happening.

The whole question of Mars sample return is a vast area of discussion and planning. Both the "forward contamination" of Mars with Earth spacecraft and the "back contamination" of Earth with return samples are serious concerns for the scientific implications of the samples and the hazards to Earth's ecology. We are getting closer to the reality of sample return, but there remains considerable work to be done before it can take place. For now, the plans consider a late 2020s mission scenario. But as with all plans that are not yet in the execution stages, one would be wise to avoid betting any money on the time frame.

HUMAN MISSIONS

"They say no plan survives first contact with implementation. I'd have to agree."

—ANDY WEIR, *The Martian*

WHEN I GIVE PRESENTATIONS ABOUT MARS the last question that invariably gets asked is: When will we send people to Mars? It is its own subject, almost divorced from the study of Mars itself. A lot gets written and hyped about the potential future missions by humans to Mars. Of course, Wernher von Braun was among the first to outline an expedition to Mars back at the dawn of the space age. There is a lot of artwork, things like people in spacesuits doing things on Mars, rockets standing upright with landing site infrastructures being assembled, piloted rovers zooming around exploring Mars, and all the other things you might imagine people doing when they arrive at Mars. There have been many science fiction stories, movies, and entertainment series about Mars expeditions and various related dramas associated with the dangerous new world.

There is so much interest in the idea of sending people to Mars that there is now a history of mock Mars missions over the last couple of decades. It seems that every year there is another team of people going off to live in a simulated Mars habitat for an extended period and exploring the human space as much as the infrastructure space of actually getting to and living on Mars. The whole topic is a source of educational activities for students in which students are given the task of designing a Mars habitat and all the related systems for living on the red planet. Again Mars almost gets lost in the discussion. The fact that you are on Mars becomes simply a prop in an expansive story of survival in a remote place. Mars science is not generally a topic of the fictional "people on Mars" storyline.

There seems to be a veritable mania with getting people on Mars. The reality is that it is a difficult task from many perspectives. The problems and technical challenges can be sorted into getting there and returning, landing there, living there, and getting around there, all while trying to stay alive if not completely healthy and sane. Books have been written on the subject and many clever concepts have been outlined.

HUMANS TO MARS

IN OCTOBER 2015 NASA HELD A conference ostensibly to begin searching for potential sites for future human Mars landings. But it was also directed at discussing the feedback between science needs for such an undertaking and the actual planning necessary for the operational requirements. In an effort to speed up the process of evaluating the requirements and generate collaboration between the various specialties that have a bearing on the whole human Mars mission concept, the workshop focused on discussing community-submitted proposals for landing sites of interest. Using the current technological understanding of the problems together with current Mars science as a guide, several ground rules were applied to site submissions.

First, the site needed to have access to places of maximum scientific value, which was assumed for purposes of discussion to be somewhere within one hundred kilometers of the landing site itself. This was defined as a region of interest, or ROI. All sorts of things were considered as scientifically interesting, but major considerations included access to deposits possibly preserving evidence of past life or past habitability, rocks of certain ages and types representative of major global geologic units, places where there were identifiable stratigraphic context and relationships, and a whole

Concept of infrastructure and
activities as humans establish a
base on Mars.

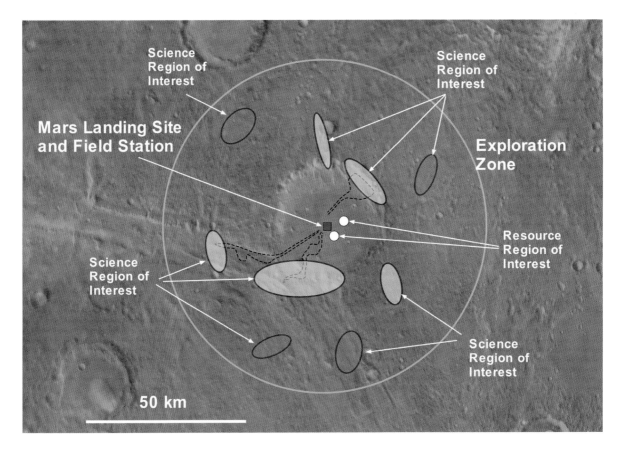

ABOVE: Depiction of the "region of interest" concept as applied to Mars human landing site selection in the latest thinking about the priorities of a human Mars mission.

list of desirables such as rocks that could include trapped atmospheric gases, volcanic rocks from known volcanic provinces, deposits from the most ancient crust, and other materials that have a high likelihood of recording things of global interest.

Second, the landing site was required to meet certain engineering needs for the actual occupation by the Mars explorers. One important need is the presence of resources that would be required for setting up habitats. This is the so-called in situ resource utilization requirement or requirement for access to sustainable extractable commodities. Then there was a civil engineering requirement that seeks a site where the necessary construction could actually be done, that is, places where the materials could be manipulated for infrastructure emplacement.

Third, the site should offer suitable potential for things like food production given the materials at hand. This simply follows from the fact that food is one of the biggest consumables and the ability to produce it as opposed to just consuming food brought from Earth offers a huge advantage in long-term sustainability.

The list goes on in more extensive details. But engaging the community in the needs of such a mission early is seen as a way of getting the necessary cross-talk between specialties that would most likely be able to provide the answers and the directions for getting those answers. More to the point, the whole process began the long conversation about the actual realities of such a mission. One reality recognized early in the process was the fact that selecting a site would no doubt require substantial verification, and that by starting now we could ensure that at least some of the initial orbital remote sensing required to proceed with evaluations could be built into current and planned missions.

WHEN WILL WE SEND HUMANS TO MARS?

A STANDARD RULE OF THUMB OVER the decades since Wernher von Braun first proposed in the late 1940s how we might send humans to Mars has been that a human mission to Mars is today's date plus twenty years, plus or minus a few years. This simply reflects the fact that it is a technologically challenging activity and that many of the unknowns are yet to be resolved. The unknowns not only include the ones I've outlined, and all the work necessary to get those needs in order, but also reflect the need to administer the process, both financially and politically. It would require international cooperation on a scale not previously done. And the funds must be available over a decade or longer without serious interruption.

Recently NASA was directed by presidential request to send humans to Mars by 2033. The NASA Transition Authorization Act of 2017 mandated that NASA "develop a human exploration roadmap . . . to expand human presence beyond low-Earth orbit to the surface of Mars and beyond, considering potential interim destinations such as cislunar space and the moons of Mars" (US Congress 2017, Section 432(b)).

This goal was analyzed in an Institute for Defense Analyses document in February 2019. The document looked at both the required technical developments and the required budgets and compared those against the current technical and budget "landscape." The document concluded "a 2033 departure date for a Mars orbital mission is infeasible under all budget scenarios and technology development and testing schedules."

Further it noted that 2035 might be possible with annual budget increases of a couple percent, but that date would be at risk from delays due to unforeseen technical problems and fabrication requiring additional work. Moreover, the development would likely impact planned lunar missions and would require less stringent "human health" considerations, in other words, would be more dangerous. By the reckoning of the Institute for Defense Analyses document, it was more likely to be 2037 before such a mission could take place and that 2039 was more realistic due to likely budget issues over the course of development. So in keeping with our rule of thumb about when a human Mars mission might occur, 2039 is about twenty years from the current date. With every year that clicks past, that date is advanced another year. And as the world recovers from the economic impact of a global pandemic, one could easily envision another decade passing before the starting gun could be fired in the race for Mars. So 2050 or beyond is sounding more possible if not probable.

Mars is a difficult place to get to and to explore. One thing that we can probably be assured of, given the complete and utter fascination that Mars has held for humanity over the course of civilization, is that we will get there. When finally we do arrive, then humans will begin the long and no doubt exciting journey of leaving tracks on the red planet on foot.

But there is one more thing we will do. Joe Aragon from Acoma related that he was involved in a thought experiment at a meeting once that asked the question: What would you want to take to a new planet where you would settle and never return? They agreed that you would want to take your mythology and build on it, creating new additions to your mythology so that it would evolve wherever you went.

EPILOGUE

IT WAS 2076 AND IT HAD been two years since they had established Hebes Base, located in Echus Chasma near the west end of the great Hebes Chasma just north of the Valles Marineris canyons. Things had finally settled down enough that the project was taking proposals from the team for small exploratory projects, mostly to shut everybody up. Being cooped up in the habitat all this time was enough to drive even the most mentally stable team member crazy. Letting members go off on worthy science exploratory projects was a good way to keep morale up. They were starting to refer to themselves as the Heebie-Jeebies, for good reason. It wasn't that they didn't ever get outside. There were plenty of trips in extravehicular activity suits to the generator located several hundred meters away. But they had been on Mars all this time, and only a couple of official trips had gone over the short horizon. Jake hadn't been on any of them. So when the opportunity came, Jake had written an exploratory proposal a year ago, and it was finally approved and stamped "GO." Now was Jake's chance to explore Mars, or kind of at least. He barely managed to get the necessary second person on board his goofy plan. Lisa was his wife and also a geologist and volcanologist like Jake, and she was also second-in-command of the habitat greenhouse. She was sick and tired of walking around in the stinky "steam house," as they called it, 24/7, sluicing garbage through hydroponic troughs and writing reports on crop yields instead of doing actual geology. This was their big chance to do some volcanology and "explore Mars" together.

On that fateful day, July 20, 2076, they had signed out and gotten signed off for use of one of the exploratory piloted helicopters for a trip to the slopes of Tharsis. The plan was to go have a look at one of those little sinkhole-like openings in the surface that were obviously skylights on deep underground caverns. Fortunately, there was a new one that was fairly small and nearby on the lower eastern reaches of the slopes around the great Tharsis Montes; a flight there would not be too taxing for the

> "Coyotes have the gift of seldom being seen; they keep to the edge of vision and beyond, loping in and out of cover on the plains and highlands. They are an old council of clowns, and they are listened to."
>
> —N. SCOTT MOMADAY, *House Made of Dawn*

helicopter. The cavi of Tharsis were well known, and Jake had argued, among other things, that it would be useful to explore the potential of such caverns for future habitats and other in situ resources. They could be a good resource in any case to have in the back pocket.

They had arrived and set up the solar panels to recharge the helicopter, deployed all the necessary safety equipment and the winch for descending into the cave opening. They agreed to take turns for safety, and Jake had won the first turn. He descended while Lisa controlled the winch and monitored things from the surface. Funny, it was the anniversary, one hundred years to the day, of the Viking 1 Lander's arrival in Chryse Planitia, looking for signs of life on Mars. In the intervening years everybody had searched high and low in all the places where water had obviously flowed and sediments had been deposited. That was the place to look, everybody said. The volcanic plateaus were dead, dry, and barren. That was the last place to look for signs of life on Mars. But here he was, in an old volcanic cavern in the most barren place on Mars. And there was life. He was excitedly communicating the discovery to Lisa, while scraping obvious microorganisms off a kind of muck from the cave walls in that last place anyone thought of looking.

Who was that old-time astrobiologist who had promoted the idea of Martian caves as abodes of life? He vaguely remembered something from a text when he was saturating his brain with Mars knowledge just before applying for the mission. Somebody named Penny Boston, or something like that. Well, she would be happy. Here it was after all. This would change a few things.

In fact, it was funny, because it was so typical of Mars discoveries. The old stories about Coyote Mars stringing humans along and doing the bait-and-switch on them was ringing true. First, it had strung them along with all those strange surface markings and, surprise, those had nothing to do with anything about Mars. Then Coyote Mars had made them dig a little deeper all those years, looking for signs of past life in simple places like old water channels and sediments. And then, surprise, that had nothing to do with the search. The evidence was there all along, but we had been looking in the wrong place. From now on it would be all about volcanic areas, baby.

BUT WAS IT THE FINAL STORY about life on Mars? Or did Coyote Mars have in mind something even more unexpected? Maybe this was just another bait waiting for that signature and shocking switch. Maybe Coyote Mars just wanted us to dig into the problem a little deeper. Then we would really be surprised.

PHOTO CREDITS

NASA/JPL/Cornell [HISTORICAL MAPS]: 154, 176-177, 180 (top), 193, 194, 196, 198-199 (top), 200-201, 205, 209, 212, 213, 214-215, 219 (bottom right), 334–335

NASA, ESA, and STScl: 8

NASA/JPL/Cornell/Texas A&M: 12 (bottom), 267 (top right, bottom two images)

Andrew Santangelo: 12 (top)

Russian Space Agency: 38 (top)

NASA/JPL: 31, 37, 38 (bottom), 52, 56, 62, 63, 83, 84, 137, 160, 173, 204, 210, 211 (side left), 219 (bottom left), 240, 247, 249 (top), 267 (top left), 291, 292

NASA/JPL-Caltech/UA: 38 (middle)

NASA/USGS: 47

Jayne Aubele: 51

NASA/JPL/Cornell/NMMNHS: 193

Author Photo (Larry Crumpler): 32, 35, 48, 65, 67, 70, 74, 75, 120, 166, 169, 174, 198-199 (bottom), 211 (top right), 217, 224 (bottom left), 249 (bottom), 316

NASA/MRO/HiRISE, LPL/ UA: 38 (center), 81, 82

NASA/USGS: 88, 91, 92 (top left)

ESA/Mars Express/HRSC/ DLR: 89

NASA/USGS/ESA/Mars Express/HRSC/DLR: 92 (middle left)

NASA/ MRO/CTX/MSSS: 93

NASA/JPL/MRO/HiRISE/ UA: 94 (top left, top middle)

NASA/USGS/MSSS/ESA/ DLR/ JPL/UA: 95 (top middle)

NASA/USGS/MRO/CTX/ MSSS: 98 (top)

NASA/JPL/MRO/HiRISE/ UA: 98 (bottom)

USGS: 83 (bottom right), 105

NWA 7034, University of New Mexico, Courtesy of Carl Agee: 119

NASA/JPL-Caltech/MRO/ HiRISE/ UA: 124, 142 (bottom right), 144, 145 (bottom left, bottom middle), 149 (top)

NASA/JPL/MSSS: 128

NASA/JPL/Mars Odyssey/ Los Alamos National Laboratory: 141

ESA/Denman productions: 142 (bottom left)

NASA/JPL-Caltech/MSSS: 145 (bottom right), 270, 271

NASA/JPL-Caltech/MSSS/ JHU-APL: 146

NASA/JPL-Caltech/ASI/UT: 147

NASA/JPL-Caltech/ University of Arizona/Texas A&M: 149 (bottom)

NASA/JPL-Caltech/ University of Arizona/Max Planck Institute: 150

NASA: 161

Drawn by Joe Aragon, Acoma Pueblo.: 167

NASA/JPL/US Geological Survey: 180 (middle left)

NASA/JPL/MRO/HiRISE/UA: 216 (top left), 241, 276 (bottom), 280

Google Earth: 216 (bottom left)

NASA/JPL/USGS/Cornell: 223, 224 (top left), 244-245, 248 (top left)

USGS/Airforce: 28

NASA/JPL-Caltech/ Cornell/ASU: 228, 230,

NASA/JPL/MRO/HiRISE/ NMMNHS: 234

NASA/JPL-Caltech/Texas A&M: 248 (top right)

NASA/JPL/Cornell/ASU/ James Sorenson/ Christian A. Lopez: 252

[NASA/JPL-Caltech/ Cornell/ Texas A&M: 263, 264

NASA/JPL-Caltech/ Cornell/Texas A&M/SSI: 267 (top right), 267 (bottom left & right)

NASA/JPL-Caltech/LANL/ CNES/ IRAP/LPGN: 273

NASA/JPL-Caltech/ASU/ UA: 276 (top), 304-305 (top left, top right, bottom right, bottom left)

NASA/ Trent Schindler: 279

NASA/JPL-Caltech: 282-283, 285, 302, 303 (bottom), 306 (top left, middle side left) 311, 314-315, 304-305 (bottom right)

NASA/JPL-Caltech/CNES/ IPGP/ Imperial College London: 284

Emirates Space Agency: 288 (top left)

China National Space Administration (CNSA): 288 (top right)

NASA/JPL/Odyssey/ THEMIS/ ASU: 289

ESA: 308 (bottom left)

JAXA/NASA: 309 (bottom right)

DLR: 308-309 (bottom middle)

NASA/JPL-Caltech/ASU/ MSSS: 300, 304-305 (bottom middle, top middle, 306 (bottom middle), 307 (bottom left, middle)

NASA/JPL-Caltech/ MRO/HiRISE/ University of Arizona: 303 (top)

RESOURCES

BOOKS

To learn even more about Mars and Mars missions, check out the books below.

Baker, David. *NASA Mars Rovers: 1997–2013 (Sojourner, Spirit, Opportunity, and Curiosity): An Insight into the Technology, History, and Development of NASA's Mars Exploration Roving Vehicles*. Sparkford, UK: Haynes Publishing, 2013.

Bell, Jim. *Postcards from Mars: The First Photographer on the Red Planet*. New York: Dutton Books, 2006.

Boyce, Joseph. *The Smithsonian Book of Mars*. Washington, DC: Smithsonian Institution Press, 2002.

Carr, M. H. *The Surface of Mars*. New Haven, CT: Yale University Press, 1981.

Cattermole, Peter. *Mars: The Story of the Red Planet*. London: Chapman and Hall, 1992.

Coles, Kenneth S., Kenneth L. Tanaka, and Philip R. Christensen. *The Atlas of Mars: Mapping Its Geography and Geology*. New York: Cambridge University Press, 2018.

David, Leonard. *Mars: Our Future on the Red Planet*. Washington, DC: National Geographic, 2016.

Godwin, Robert. *Mars: The NASA Mission Reports (2 volumes)*. Burlington, Ontario, Canada: Apogee Books, 2005.

Greeley, Ronald. *Planetary Landscapes*. London: Allen and Unwin, 1985.

Hartmann, William K. *A Traveler's Guide to Mars: The Mysterious Landscapes of the Red Planet*. New York: Workman, 2003.

Kieffer, Hugh, Bruce Jakosky, Conway Snyder, and Mildred Matthews, eds. *Mars*. Tucson: University of Arizona Press, 1992.

Manning, Rob. *Mars Rover Curiosity: An Inside Account from Curiosity's Chief Engineer*. Washington, DC: Smithsonian Books, 2014.

McEwen, Alfred S., Candice Hansen-Koharcheck, and Ari Espinoza. *Mars: The Pristine Beauty of the Red Planet*. Tucson: University of Arizona Press, 2017.

Moore, Patrick. *On Mars*. London: Cassell, 1998.

Rusch, Elizabeth. *The Mighty Mars Rovers: The Incredible Adventures of Spirit and Opportunity*. Boston: Houghton Mifflin Books, 2012.

Sawyer, Kathy. *The Rock from Mars: A Detective Story on Two Planets*. New York: Random House, 2006.

Squyres, Steve. *Roving Mars: Spirit, Opportunity, and the Exploration of the Red Planet*. New York: Hachette, 2005.

Wiens, Roger. *Red Rover: Inside the Story of Robotic Space Exploration, from Genesis to the Mars Rover Curiosity*. New York: Basic Books, 2013.

WEBSITES

To see more amazing pictures of Mars missions, check out these websites:

NASA Photojournal
https://photojournal.jpl.nasa.gov

Mars Odyssey THEMIS
https://themis.mars.asu.edu/gallery

Mars Reconnaissance Orbiter
https://mars.nasa.gov/mro/

https://www.nasa.gov/mission_pages
/MRO/

MRO/HiRISE Images
https://www.uahirise.org

Spirit and Opportunity
https://www.nasa.gov/mission_pages
/mer/

**Spirit and Opportunity Images/
Data through MER Analyst's
Notebook**
https://an.rsl.wustl.edu/mer

Curiosity Rover
https://www.nasa.gov/mission_pages
/msl/

InSight Mars Lander
https://www.nasa.gov/mission_pages
/insight/main/index.html

Perseverance. Mars Rover
https://www.nasa.gov/perseverance

MAPS

If you want to see more maps of Mars, take a look at these resources:

Mars Geologic Map
Tanaka, Kenneth L., James A. Skinner, James M. Dohm, et al. 2014. *Geologic Map of Mars.* 1:20,000,000 scale. US Geological Survey, Scientific Investigations Map 3292.

https://dx.doi.org/10.3133/sim3292;
https://www.usgs.gov/media/images
/geologic-map-mars

**Mars 2020/Perseverance
Landing Site**
Sun, Vivian Z., and Kathryn M. Stack. 2020. *Geologic Map of Jezero Crater and the Nili Planum Region, Mars.* 1 sheet, 1:75,000 scale. US Geological Survey, Scientific Investigations Map 3464.

https://doi.org/10.3133/sim3464

ACKNOWLEDGMENTS

THIS BOOK WAS WRITTEN DURING A global pandemic, which frankly provided long hours of isolated, hermit-like enjoyment in the writing process, interrupted only by pleasant hikes up the canyon next to the house to reflect on the Mars-like geology of New Mexico. I have to say that after several months of seeing other people only on screens, I felt like I was some colonist on Mars communicating with an Earth that was far away.

One thing you learn when writing a book like this is that many of the most important stories and people are missing in the final manuscript, not because you forgot to include them, but because they are not included due to vagaries of writing, the arc of the narrative, and the editors' experience in keeping the book from becoming way too long. It is pretty easy to recall the people who have been important along the professional road. As usual, it is a list of people who played a mentor role in one fashion or another. I was the first person in my family to go to college. My father was a truck mechanic and my mother was a 50s-era housewife. Neither had any idea why I wanted a telescope, but they scraped together money to buy me one and later support my college efforts. Charles Welby, my undergraduate academic advisor at NC State University, provided a solid background in the fundamentals of geology and supported my odd fascination with planetary things. Wolf Elston at the University of New Mexico did that and more by ensuring I had real experience with real geology, and met all the leaders in the field, and supported both my planetary and field geology goals. Later, Jim Head at Brown University provided a lot of philosophical and practical insight and opportunities into the world of professional big science. I am pleased that there is now a crater on Mars named after Wolf on the edge of Hesperia Planum.

Those are a few of the principal people who helped keep me going along the professional road. Then there are all the people who supported me in one way or another who I do not know. That is a reference of course to the reviewers, often anonymous, of professional science manuscripts and proposals who supported whatever research or proposal I had submitted. We all know that reviews of one's important research efforts are somewhat a crapshoot. Writing a good paper or proposal is just a permit to roll the dice. Then the reviewers may or may not like the manuscript according to their background. You can get good reviews and bad reviews from the same manuscript depending on the reviewers. If the balance is more one way or the other, outcomes are greatly changed. And it is a significant factor in success not recognized often enough. So this is a thank you to all those who provided supportive reviews.

During my career, I moved from the university world to the museum world, and I am pleased that I had the opportunity to learn the techniques, and now have the enjoyment of communicating science to nonscientists.

As I entered this new world of book publishing, a big thank you for support and advice from New Mexico colleagues including science history author Loretta Hall, science fiction author Joan Saberhagen, and writer and editor John Byram.

And to my wife, Jayne Aubele, who read sections, chapters, and the entire manuscript several times and offered invaluable edits and suggestions from the perspective of a fellow planetary geologist.

Rachel Tillman, who started the Viking Mars Mission Education and History Project, and made contact with both Jayne and me to interview us about our experiences, helped to focus my thinking about the historical impact of that amazing mission.

No book manuscript can happen without good editors and I certainly had that. Myles Archibald (London office), and Lisa Sharkey and Maddie Pillari (New York office), patiently guided me through the process of getting a book manuscript where it needed to be and when it needed to be there. Surely, they have done this with many authors a hundred times before, but their patience and engagement did not betray that at all. It was a pleasure working with professionals who are so good at what they do. It made my job of writing easy and fun.

Our cats, Akna and Freyja, were endless sources of companionship and joy during the many years of Mars research and mission work. They were adopted during the Magellan mission to Venus, and named for surface features on that planet. Sadly, both died during the years of the Spirit and Opportunity missions. They actually enjoyed traveling; and, someday, I am sure that companion cats will take full advantage of climbing and jumping in a warm, cozy habitat in the low gravity of Mars.

INDEX

Page references in *italics* indicate images.

Smith, Peter H., 148

Smithsonian Air and Space Museum, 135, 168

SNCs, 119–20

Soderblom, Larry, 55, 106

Solis Planum, 83, *83*

southern hemisphere, 40, 73, 76, 82

Soviet Union, 30, 37, 41, 125, 126, 130

Spirit Rover, 11, 143, 256, 257, 258, 262, *263*, 265, 266, *267*, 296, 308
  Adirondack and, 195
  arrival at Mars, 168–74, *167*, 168, *169*, *174*, 177, *176–7*, 185
  Bonneville Crater, 197, *199*, 222
  collection of tracks left by, *196*
  Columbia Hills and, 176, 189, 192, 193–4, 202–3, *204*, 207–8, 209–15, *210*, *211*, 212, *212*, 213, *213*, 223, 226, 233–4
  death of, 243–8, *245*
  design lifetime, 190–2, 193
  "drama queen," 178, 179
  dust devils of Mars and, 240–1, *240*, 242
  first panorama showing view south from landing sire, *176–7*, 177
  first significant drives of, 186, 187
  flat tire, 239
  Gusev Crater and, 179, 181, 185, 192–3, 226, 233
  Home Plate, 216–17, *216*, 230–1, 232
  Mars Express and, 141
  Martian night sky and, 265–6, *267*
  naming of, 140, 160–1, *161*, 167, *167*
  RAT grinder on, 237

Route 66 and, 223, 224, *224*

"selfies" taken by, comparison of, *209*

sol 18 anomaly, 178–9, 195

sol 63 image of the morning sky taken by, *12*

solar panels, 191, 262

Tantalus Crater and, 243–5

undae and, 99

Squyres, Steve, 129, 159, 179, 189, 191, 249, *249*

Stegner, Wallace ("Thoughts on a Dry Land"), 94

Strode, Muriel ("Wind-Wafted Wild Flowers"), 191

sulcus, 79

Sullivan, Anne, 254

Sullivan, Rob, 219–20

Sunset Crater, 43

Syrtis Major Planum, 83

T

Tanaka, Ken, 108–9

Tantalus Crater, 243, 245

tau, 242, 243, 250, 262–3, *263*

Taylor, Jeff, 132

TERMOSKAN, 126

terra, 78, 84–5

Tharsis region, 69, 71, 83, 88, *88*, 90–1, 92, *92*, 93, *93*, 112, 145, *145*, 319–20

Theiikian Eon, 113, 114, 277

Thermal Emission Spectrometer (TES), 128
  Mini-TES instrument, 140, 223–4, 231, 237

Thermal Imaging Spectrometer (Thermal Emission Imaging System, or THEMIS), 32, *32*, 140, 145, *289*

tholus, 78

Tianwen-1, 289, *289*

time on Mars, 257–61, *259*

Tractus Fossae, 92, *92*, 93, *93*

Tsiolkovsky, Konstantin, 30

Twain, Mark, 29

U

UAE Hope (Emirates Mars Mission Orbiter), 287–8, *288*

unconformity, 95–6, 201

Unda, 78, 96–7, *98*, 99–100

University of Arizona, 52, 123, 147, 148

University of New Mexico, 44, 120, *120*, 123, 272

US Air Force map of Mars, 27–8, *28*, 48

US Geological Survey, 43, 55, 106, 108, 136

Utopia Planitia, 56–7, 61, 62, *62*, 63, 87, 289, *289*

V

Vallejos, Mark, 166, *166*

valles, 78, 93–4

Valles Marineris, 40, 69, 91, *91*, 92, *92*, 319

vallis, 78

vastitas, 78

vastitas Borealis, 99

Venus, 116, 119, 120–1, 123, 136, 327

Vernadsky Institute, 123

Spirit's view of the surroundings in
the Columbia Hills on Sols 410 to 413
(February 27 to March 2, 2005)
from a position known informally
as "Larry's Lookout."

ABOUT THE AUTHOR

DR. LARRY CRUMPLER IS CURRENTLY THE research curator of Volcanology and Space Science at the Museum of Natural History and Science. He is both a planetary geologist and a field geologist, specializing in volcanic geology of Mars and Earth and geology of the Southwest. He has authored many research papers and popular articles on planetary and Earth geology. He is a Fellow of the Geological Society of America (GSA) and former chair of the Planetary Geology Division of GSA. He has participated in many NASA planetary missions over the past several decades including the Viking missions, Venus Magellan mission, Mars Exploration Rovers (Spirit and Opportunity) mission, Mars Reconnaissance Orbiter mission, and Mars 2020/Perseverance rover mission. As the long-term planning lead for the Mars Exploration Rover Program, Dr. Crumpler was part of the daily telecons that defined the ongoing activities of the rovers on Mars for more than fourteen years. A location on the planet—"Larry's Lookout"—was informally named after Dr. Crumpler. He lives in Albuquerque with his wife, Jayne, and two cats, where they explore the Mars-like geology of New Mexico as often as possible, and he maintains and shows several vintage vehicles, and writes about, photographs, and does research on New Mexico's geology and its one thousand volcanoes.